Design and Development of Radio Frequency Identification (RFID) and RFID-Enabled Sensors on Flexible Low Cost Substrates

Synthesis Lectures on RF/Microwaves

Editor
Amir Mortazawi, *University of Michigan*

Design and Development of Radio Frequency Identification (RFID) and RFID-Enabled Sensors on Flexible Low Cost Substrates
Li Yang, Amin Rida, and Manos M. Tentzeris
2009

Design and Development of Radio Frequency Identification (RFID) and RFID-Enabled Sensors on Flexible Low Cost Substrates

Li Yang, Amin Rida, and Manos M. Tentzeris

ISBN: 978-3-031-01396-6   paperback
ISBN: 978-3-031-02524-2   ebook

DOI 10.1007/978-3-031-02524-2

A Publication in the Springer series
Synthesis Lectures on RF/Microwaves

Lecture #1
Series Editor: Amir Mortazawi, *University of Michigan*

Series ISSN
Synthesis Lectures on RF/Microwaves
ISSN pending.

Design and Development of Radio Frequency Identification (RFID) and RFID-Enabled Sensors on Flexible Low Cost Substrates

Li Yang, Amin Rida, and Manos M. Tentzeris
Georgia Institute of Technology

SYNTHESIS LECTURES ON RF/MICROWAVES #1

ABSTRACT

This book presents a step-by-step discussion of the Design and Development of Radio Frequency Identification (RFID) and RFID-enabled Sensors on Flexible Low Cost Substrates for the UHF Frequency bands. Various examples of fully function building blocks (design and fabrication of antennas, integration with ICs and microcontrollers, power sources, as well as inkjet-printing techniques) demonstrate the revolutionary effect of this approach in low cost RFID and RFID-enabled sensors fields. This approach could be easily extended to other microwave and wireless applications as well. The first chapter describes the basic functionality and the physical and IT-related principles underlying RFID and sensors technology. Chapter two explains in detail inkjet-printing technology providing the characterization of the conductive ink, which consists of nano-silver-particles, while highlighting the importance of this technology as a fast and simple fabrication technique especially on flexible organic substrates such as Liquid Crystal Polymer (LCP) or paper-based substrates. Chapter three demonstrates several compact inkjet-printed UHF RFID antennas using antenna matching techniques to match IC's complex impedance as prototypes to provide the proof of concept of this technology. Chapter four discusses the benefits of using conformal magnetic material as a substrate for miniaturized high-frequency circuit applications. In addition, in Chapter five, the authors also touch up the state-of-the-art area of fully-integrated wireless sensor modules on organic substrates and show the first ever 2D sensor integration with an RFID tag module on paper, as well as the possibility of 3D multilayer paper-based RF/microwave structures.

The authors would like to express our gratitude to the individuals and organizations that helped in one way or another to produce this book. First to the colleagues in ATHENA research group in Georgia Institute of Technology, for their contribution in the research projects. To the staff members in Georgia Electronic Design Center, for their valuable help. To Jiexin Li, for her continuous support and patience. To Amir Mortazawi, our series editor, for his guidance. Also, the book would not have been developed without the very capable assistance from Joel D. Claypool, and other publishing professionals at Morgan & Claypool Publishers.

KEYWORDS

RFID, RFID-enabled Sensor, UHF, Conformal antennas, Matching techniques, Inkjet printing, Flexible substrate, Organic substrate, Conformal magnetic composite, Printable electronics

Contents

CHAPTER 1

Radio Frequency Identification Introduction

1.1 HISTORY OF RADIO FREQUENCY IDENTIFICATION (RFID)

Radio Frequency Identification (RFID) is a rapidly developing automatic wireless data-collection technology with a long history. The first multi-bit functional passive RFID systems, with a range of several meters, appeared in the early 1970s, and continued to evolve through the 1980s. Recently, RFID has experienced a tremendous growth, due to developments in integrated circuits and radios, and due to increased interest from the retail industrial and government. Thus, the first decade of the 21st century sees the world moving toward the technology's widespread and large-scale adoption. A major landmark was the announcement by Wal-Mart Inc. to mandate RFID for its suppliers in "the near future," at the Retail Systems Conference in June 2003 in Chicago. This was followed by the release of the first EPCglobal standard in January 2005. It has been predicted that worldwide revenue for RFID will eclipse $1.2 billion in 2008, marking an almost 31% increase over the previous year [1]. Key volume applications for RFID technology have been in markets such as access control, sensors and metering applications, payment systems, communication and transportation, parcel and document tracking, distribution logistics, automotive systems, livestock/pet tracking, and hospitals/pharmaceutical applications [2].

An RFID system consists of readers and tags. A typical system has a few readers, either stationary or mobile, and many tags which are attached to objects. The near-field and far-field RFID coupling mechanisms are shown in Fig. 1.1. A reader communicates with the tags in its wireless range and collects information about the objects to which tags are attached. RFID technology has brought many advantages over the existing barcode technology. RFID tags can be embedded in an item rather than the physical exposure requirement of barcodes and can be detected using radio frequency (RF) signal. The communication based on RF signal also enhances the read range for RFID tags. In addition, barcodes only contain information about the manufacturer of an item and basic information about the object itself; however, RFID is particularly useful for applications in which the item must be identified uniquely. RFID also can hold additional functionality which means more bits of information.

The roots of RFID technology can be traced back to World War II. Both sides of the war were using radar to warn of approaching planes while they were still miles away; however, it was impossible to distinguish enemy planes from allied ones. The Germans discovered that by just rolling planes when returning to base changes the radio signal reflected back which would alert the radar

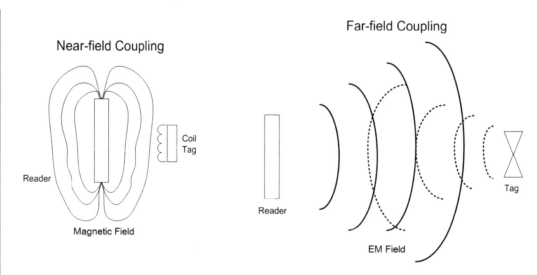

Figure 1.1: Near-field and far-field RFID coupling mechanisms.

crew on the ground. This crude method made it possible for the Germans to identify their planes. The British developed the first active identify friend or foe (IFF) system. By just putting a transmitter on each British plane, it received signals from the aircraft and identified it as a friend [3].

An early exploration of the RFID technology came in October 1948 by Harry Stockman [4]. He stated back then that "considerable research and development work has to be done before the remaining basic problems in reflected-power communication are solved, and before the field of useful applications is explored." His vision flourished until other developments in the transistor, the integrated circuit, the microprocessor, and the communication networks took place. RFID had to wait for a while to be realized [5].

The advances in radar and RF communications systems continued after World War II through the 1950s and 1960s, as described in Table 1.1. In 1960s application field trials initiated. The first commercial product came. Companies were investigating solutions for anti-theft and this revolutionized the whole RFID industry. They investigated the anti-theft systems that utilized RF waves to monitor if an item is paid or not. This was the start of the 1-bit Electronic Article Surveillance (EAS) tags by Sensormatic, Checkpoint, and Knogo. This is by far the most commonly used RFID application.

The electronic identification of items caught the interest of large companies as well. In 1970s large corporations like Raytheon (RayTag 1973), RCA, and Fairchild (Electronic Identification system 1975, electronic license plate for motor vehicles 1977) built their own RFID modules. Thomas Meyers and Ashley Leigh of Fairchild also developed a passive encoding microwave transponder in 1978 [5].

Table 1.1: The Decades of RFID

Decade	Event
1940-1950	Radar refined and used, major World War II development effort. RFID invented in 1948.
1950-1960	Early explorations of RFID technology, laboratory experiments.
1960-1970	Development of the theory of RFID. Start of applications field trials.
1970-1980	Explosion of RFID development. Tests of RFID accelerate. Very early adopter implementations of RFID.
1980-1990	Commercial applications of RFID enter mainstream.
1990-2000	Emergence of standards. RFID widely deployed. RFID becomes a part of everyday life.

By 1980s there were mainstream applications all around the world. The RFID was like a wildfire spreading without any boundaries. In the United States, RFID technology found its place in transportation (highway tolls) and personnel access (smart ID cards). In Europe, short-range animal tracking, industrial and business systems RFID applications attracted the industry. Using RFID technology, world's first commercial application for collecting tolls in Norway (1987) and after in the United States by the Dallas North Turnpike (1989) were established.

In 1990s, IBM engineers developed and patented a UHF RFID system. IBM conducted early research with Wal-Mart, but this technology was never commercialized. UHF offered longer read range and faster data transfer compared to the 125 kHz and 13.56 MHz applications. With these accomplishments, it led the way to the world's first open highway electronic tolling system in Oklahoma in 1991. This was followed by the world's first combined toll collection and traffic management system in Houston by the Harris County Toll Road Authority (1992). In addition to this, GA 400 and Kansas Turnpike Highways were the first to implement multi-protocol tags which allowed two different standards to be read [3, 5].

After IBM's early pilot studies in 1990s with Wal-Mart, UHF RFID got a boost in 1999, when the Uniform Code Council, European Article Number (EAN) International, Procter & Gamble and Gillette teamed up to establish the Auto-ID Center at the Massachusetts Institute of Technology. This research focused on putting a serial number on the tag to keep the price down using a microchip and an antenna. By storing this information in a database, tag tracking was finally realized in this grand networking technology. This was a crucial point in terms of business because now a stronger communication link between the manufacturers and the business partners was established. A business partner would now know when a shipment was leaving the dock at a manufacturing facility or warehouse, and a retailer could automatically let the manufacturer know when the goods arrived [3].

The Auto-ID Center also initiated the two air interface protocols (Class 1 and Class 0), the Electronic Product Code (EPC) numbering scheme, and the network architecture used to seek for the RFID tag data between 1999 and 2003. The Uniform Code Council licensed this technology in 2003 and EPCglobal was born as a joint venture with EAN International, to commercialize EPC technology.

Today some of the biggest retailers in the world such as Albertsons, Metro, Target, Tesco, Wal-Mart, and the U.S. Department of Defense stated that they plan to use EPC technology to track their goods. The healthcare/pharmaceutical, automotive, and other industries are also pushing towards adaptation of this new technology. EPCglobal adopted a second generation (Gen-2 ISO 18000-6-C) standard in January 2005. This standard is widely used in the RFID world today [3].

1.2   CHALLENGES IN RFID TAG DESIGN

For a successful RFID implementation one has to possess a keen knowledge of its standards, its technology, and how it meets the different needs for various applications. FedEx CIO Rob Carter quoted Bill Gates' definition of a "2-10 technology" in an interview when he was asked about RFID. "2-10 technology" means for the first two years, hype reigns, followed by disappointment, until the day 10 years later when people realize the technology has flourished and become part of the daily life. Carter accepts after noticing some challenges and problems FedEx is experiencing with tags, "RFID might be a 3-15 technology." [6]. This citing comes from a man who is in charge of the whole activity of tracking parcels it does not even own for up to 48 hours anywhere in the world – an activity that cries out for RFID.

Apart form higher level problems in RFID applications, tag design imposes different lower level challenges. These challenges include current high cost of tags, tag performance issues, and integration with sensors for sensing capabilities. From a system point of view problems at the lower level must be resolved before moving up on the RFID system hierarchy for an optimized overall performance.

1.2.1   THE COST OF RFID TAG

In order to sell RFID tags just like any other product it has to be cheap. RFID is intended to produce an electronic replacement for the ubiquitous UPC barcode. By implementing the barcode in electronic form, it is expected that item-level RFID will enable automated inventory control in supermarkets and department stores, will facilitate rapid checkout, and will also allow more efficient product flow from the manufacturer to the consumer with reduced overall wastage and idle inventory. Individually tagged items typically have a price floor in the range of a few cents to few tens of cents. Given typical price margins, it will therefore be necessary to deliver a tag with a total price perturbation of perhaps less than one cent to allow widespread deployment [7]. In contrast, pallet-level tracking solutions that are currently being deployed have price-points larger than ten cents. Mark Roberti's report [8] based on Auto-ID Center's predictions on IC manufacturing cost reduction [9] indicates that in the near future the cost of a passive tag can reach as low as 5 cents from 30-35 cents [10] as it is now. The prediction relies on the fact that these tags will be sold in high volume about 30 billion a year which would in return reduce the cost of ICs to almost 1 cent. The rest of the cost will be distributed in the cost of substrate and the assembly process. Paper-based substrate is a promising candidate for the low-cost substrate material. The high demand and the mass production of paper make it widely available and the lowest cost material ever made [11]. Using paper as the substrate for RFID tags can dramatically reduce the material cost. However, there are hundreds of different paper materials available in the commercial market, varying in density, coating, thickness, texture, etc. Each has its own RF characteristics. Therefore, the RF characterization of paper substrate becomes a must for optimal designs utilizing this low-cost substrate. Some characterization work has been done in frequencies beneath UHF band [12, 13, 14], but none – to the authors' knowledge – in or above UHF band. No paper-based RFID tag has been reported either.

1.2.2   TAG PERFORMANCE

Tag performance in an RFID system is mostly evaluated by how the tag read range is in different environments. This depends mainly on the tag IC and antenna properties as well as the propagation environment. The tag characteristics can be summed up in IC sensitivity, antenna gain, antenna polarization, and impedance match. The propagation environment limitations are the path loss and tag detuning [15].

Unlike most of the other RF front-ends in which antennas have been designed primarily to match either 50Ω or 75Ω loads for years, RFID tag antenna has to be directly matched to the IC chip which primarily exhibits complex input impedance. This is because in order to maximize the performance of the transponder, maximum power must be delivered from the antenna to the IC. Therefore, impedance matching technique plays an important role in a successful RFID tag design.

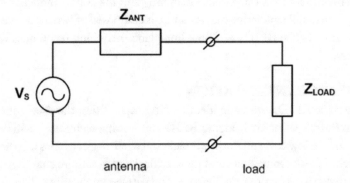

Figure 1.2: The equivalent circuit of an RFID tag.

The equivalent circuit of the antenna-load is shown in Fig. 1.2. V_s is the voltage across the antenna, which is induced from the receiving signal. The antenna displays complex input impedance Z_{ANT} at its terminals. The chip also displays complex impedance Z_{LOAD}, when looking into the opposite direction of the antenna. The load's impedance is depended on the IC and can be measured.

In order to ensure maximum power transfer from the antenna to the load, the input impedance of the antenna must be conjugately matched to the IC's impedance in the operating frequency of the tag [16], as depicted in Equation (1.1). In other words, the real part of the antenna input impedance must be equal to the real part of the load's impedance and the imaginary part of the antenna input impedance must be equal to the opposite of the imaginary part of the load's impedance [17].

$$Z_{ANT} = Z_{LOAD}^* \qquad (1.1)$$

Kurokawa [18] described a concept of power waves traveling between the generator and load, and introduced the following definitions for the power reflection coefficient $|s|^2$, as shown in Equation (1.2).

$$|s|^2 = \left| \frac{Z_{LOAD} - Z_{ANT}^*}{Z_{LOAD} + Z_{ANT}} \right|^2, \qquad 0 \le |s|^2 \le 1 \qquad (1.2)$$

The power reflection coefficient $|s|^2$ shows what fraction of the maximum power available from the antenna is not delivered to the load [19]. As a result, achieving maximum power transfer from the antenna to the load is translated into minimizing the power reflection coefficient $|s|^2$. It has to be noted that both the impedance of the antenna and the load vary with frequency. For this reason $|s|^2$ can be minimized in a single frequency. Consequently, this is chosen to be the operation frequency of the RFID tag.

Adding an external matching network with lumped elements is usually prohibited due to cost, fabrication and size issues. Instead, serial stub feed structure has been proved an effective method for the impedance match, as illustrated in Fig. 1.3 [20]. The resistive shorting stub and the double inductive stub make up the overall matching network to match to the chip input impedance. The shorting stub mainly controls the resistive matching and the double inductive stub controls the reactive matching. The double inductive stub structure is composed of two inductive stubs to provide symmetry on both sides of the RFID tag. More impedance matching techniques will be illustrated in the later chapters.

1.2.3   RFID/SENSOR INTEGRATION

In addition to the basic RFID automatic identification capabilities, the capabilities of integrating wireless sensors on flexible substrate bridging RFID and sensing technology will be demonstrated.

As the demand for low cost and flexible broadband wireless electronics increases, the materials and integration techniques become more and more critical and face more challenges, especially with the ever growing interest for "cognitive intelligence" and wireless applications. This demand is further enhanced by the need for inexpensive, reliable, and durable wireless RFID-enabled sensor nodes that is driven by several applications, such as logistics, anti-counterfeiting, supply-chain monitoring, healthcare, pharmaceutical, and is regarded as one of the most disruptive technologies to realize truly ubiquitous ad-hoc networks. The aim is to create a system that is capable of not only tracking, but also monitoring. With this real-time cognition, the status of a certain object will be made possible by a simple function of a sensor integrated in the RFID tag, achieving the ultimate goal of

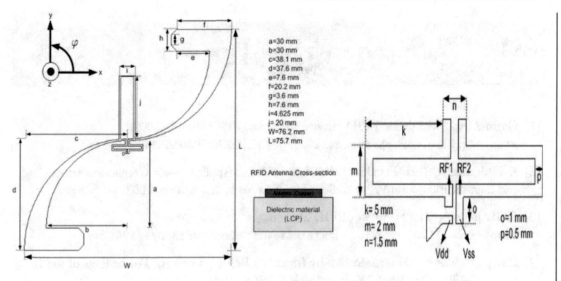

Figure 1.3: RFID tag antenna with serial stub feed structure for impedance matching [20].

creating a secured "intelligent network of RFID-enabled sensors." Design considerations including RFID/sensor interface and power consumption issue will be addressed, accompanied with design prototype examples.

Bibliography

[1] Gartner Inc. "Worldwide RFID revenue to surpass $1.2 billion in 2008,"
http://www.mb.com.ph/issues/2008/03/05/TECH20080305118642.html

[2] K. Finkenzeller, RFID Handbook: Fundamentals and Applications in Contactless Smart Cards and Identification, John Wiley & Sons Inc, New York, 2nd edition, 2003.

[3] RFID Journal, "The History of RFID Technology,"
http://www.rfidjournal.com/article/articleview/1338/1/129

[4] Harry Stockman, "Communication by Means of Reflected Power," Proceedings of the IRE, pp. 1196–1204, Oct. 1948. DOI: 10.1109/JRPROC.1948.226245

[5] Jeremy Landt, "Shrouds of Time The History of RFID," AIM Inc., ver. 1.0. Oct. 2001. DOI: 10.1109/MP.2005.1549751

[6] Howard Baldwin, "How to Handle RFID's Real-world Challenges," Microsoft Corporation, 2006.
http://www.microsoft.com/midsizebusiness/businessvalue/rfidchallenges.mspx

[7] V. Subramanian, and J. Frechet, "Progress toward development of all-printed RFID tags: materials, processes, and devices," Proceedings of the IEEE, vol. 93, no. 7, pp. 1330–1338, July 2005. DOI: 10.1109/JPROC.2005.850305

[8] Mark Roberti, "Tag Cost and ROI,"
www.rfidjournal.com/article/articleview/796/

[9] Gitanjali Swamy, and Sanjay Sarma, "Manufacturing Cost Simulations for Low Cost RFID Systems," Auto-ID Center, Feb. 2003.

[10] Tracking RFID: What Savvy IT Managers Need to Know Today,
www.sun.com/emrkt/innercircle/newsletter/0305cto.html

[11] Antonio Ferrer-Vidal, Amin Rida, Serkan Basat, Li Yang, and Manos M. Tentzeris, "Integration of Sensors and RFID's on Ultra-low-cost Paper-based Substrates for Wireless Sensor Networks Applications," Wireless Mesh Networks, 2006. WiMesh 2006. 2nd IEEE Workshop on, pp. 126–128, Reston, VA, 2006. DOI: 10.1109/WIMESH.2006.288610

[12] S. Simula, S. Ikalainen, and K. Niskanen, "Measurement of the Dielectric Properties of Paper," Journal of Imaging Science and Tech. Vol. 43, No. 5, September 1999.

[13] H. Ichimura, A. Kakimoto, and B. Ichijo, "Dielectric Property Measurement of Insulating Paper by the Gap Variation Method," IEEE Trans. Parts, Materials and Packaging, Vol. PMP-4, No. 2, June, 1968. DOI: 10.1109/TPMP.1968.1135885

[14] L. Apekis, C. Christodoulides, and P. Pissis, "Dielectric properties of paper as a function of moisture content," Dielectric Materials, Measurements and Applications, 1988., Fifth International Conference, pp. 97–100, 27-30 Jun 1988.

[15] P. V. Nikitin, and K. V. S. Rao, "Performance Limitations of Passive UHF RFID Systems," IEEE Antennas and Propagation Society Symp., pp. 1011–1014, July 2006. DOI: 10.1109/APS.2006.1710704

[16] K. V. S. Rao, Pavel V. Nikitin, and S. F. Lam, "Impedance Matching Concepts in RFID Transponder Design," Fourth IEEE Workshop on Automatic Identification Advanced Technologies, AutoID'05, pp. 39–42, 2005. DOI: 10.1109/AUTOID.2005.35

[17] David M. Pozar, Microwave Engineering, 3rd Edition, John Wiley & Sons Inc., 2005.

[18] K. Kurokawa, "Power Waves and the Scattering Matrix," Microwave Theory and Techniques, IEEE Transactions on., vol. MTT-13, no. 3, pp. 194–202, Mar. 1965. DOI: 10.1109/TMTT.1965.1125964

[19] P. V. Nikitin, K. V. S. Rao, S. F. Lam, V. Pillai, R. Martinez, and H. Heinrich, "Power Reflection Coefficient Analysis for Complex Impedances in RFID Tag Design," IEEE Transactions on Microwave Theory and Techniques, vol. 53, issue 9, pp. 2721–2725, 2005. DOI: 10.1109/TMTT.2005.854191

[20] S. Basat, S. Bhattacharya, L. Yang, A. Rida, M. M. Tentzeris, and J. Laskar, "Design of a Novel High-efficiency UHF RFID Antenna on Flexible LCP Substrate with High Read-Range Capability," Procs. of the 2006 IEEE-APS Symposium, pp. 1031–1034, Albuquerque, NM, July 2006. DOI: 10.1109/APS.2006.1710709

CHAPTER 2

Flexible Organic Low Cost Substrates

2.1 PAPER: THE ULTIMATE SOLUTION FOR LOWEST COST ENVIRONMENTALLY FRIENDLY RF SUBSTRATE

There are many aspects of paper that make it an excellent candidate for an extremely low-cost substrate for RFID and other RF applications. Paper; an organic-based substrate, is widely available; the high demand and the mass production of paper make it the cheapest material ever made. From a manufacturing point of view, paper is well suited for reel-to-reel processing, as shown in Fig. 2.1, thus mass fabricating RFID inlays on paper becomes more feasible. Paper also has low surface profile and, with appropriate coating, it is suitable for fast printing processes such as direct write methodologies instead of the traditional metal etching techniques. A fast process, like inkjet printing, can be used efficiently to print electronics on/in paper substrates. This also enables components such as: antennas, IC, memory, batteries and/or sensors to be easily embedded in/on paper modules. In addition, paper can be made hydrophobic as shown in Fig. 2.2, and/or fire-retardant by adding certain textiles to it, which easily resolve any moisture absorbing issues that fiber-based materials such as paper suffer from [1]. Last, but not least, paper is one of the most environmentally-friendly materials and the proposed approach could potentially set the foundation for the first generation of truly "green" RF electronics and modules.

However; due to the wide availability of different types of paper that varies in density, coating, thickness, and texture, dielectric properties: dielectric constant and dielectric loss tangent, or dielectric RF characterization of paper substrates becomes an essential step before any RF "on-paper" designs. The electrical characterization of paper need to be performed and results have shown the feasibility of the use of paper in the UHF and RF frequencies.

Another note to mention here is that the low cost fabrication and even the assembly with PCB compatible processes can realize paper boards similar to printed wiring boards, which can support passives, wirings, RFID, sensors, and other components in a 3D multi-layer platform [2]–[8].

2.2 DIELECTRIC CHARACTERIZATION OF THE PAPER SUBSTRATE

RF characterization of paper becomes a critical step for the qualification of the paper material for a wide range of frequency domain applications. The knowledge of the dielectric properties such as dielectric constant (ε_r) and loss tangent ($\tan \delta$) become necessary for the design of any high frequency structure such as RFID antennas on the paper substrate and more importantly if it is to

Figure 2.1: Reels of paper.

Figure 2.2: Magnified droplet of water sitting on a paper substrate.

be embedded inside the substrate. Precise methods for high-frequency dielectric characterization include microstrip ring resonators, parallel plate resonators, and cavity resonators [9]. In an extensive literature review, such properties were not found to be available for paper for the desired application frequency range (above 900 MHz).

In order to measure the dielectric constant (ε_r) and loss tangent (tan δ) of paper up to 2 GHz, a microstrip ring resonator structure was designed; the configuration diagram is shown in Fig. 2.3. A calibration method namely through-reflect-lines (TRL) was utilized to de-embed the effect of the feeding lines. It is to be noted that tan δ extraction using the microstrip ring resonator approach requires reliable theoretical equations for the estimation of the conductor losses [10].

Among the critical needs for the selection of the right type of paper for electronics applications are the surface planarity, water-repelling, lamination capability for 3D module development, via-forming ability, adhesion, and co-processability with low-cost manufacturing. For the trial runs, a commercially available paper with hydrophobic coating was selected. The thickness of the single sheet of paper is 260±3 μm. An 18 μm thick copper foil was selected as the metallic material heat-bonded on both sides of the paper substrate, in order to accurately model and de-embed the conductive loss of the microstrip circuit. The photolithography process was conducted using a dry film photo-resist followed by UV exposure and finally etching copper using a slow etching methodology. The paper substrate was then dried at 100°C for 30 minutes.

To investigate the sensitivity of the results to the paper thickness as well as to investigate the effect of the bonding process, 9 sheets of paper were directly heat-bonded together to grow a thickness of 2.3 mm, without any extra adhesive layers.

The characterization covers the UHF RFID frequency band that is utilized by applications that are commonly used in port security, inventory tracking, airport security and baggage control, automotive and pharmaceutical/healthcare industries.

The ring resonator produces Insertion Loss (S_{21}) results with periodic frequency resonances. In this method, ε_r can be extracted from the location of the resonances of a given radius ring resonator while tan δ is extracted from the quality factor (Q) of the resonance peaks along with the theoretical calculations of the conductor losses. Measurements of S_{21} were done over the frequency range 0.4 GHz to 1.9 GHz using Agilent 8530A Vector Network Analyzer (VNA). Typical SMA coaxial connectors were used to feed the ring resonator structure. TRL calibration was performed to de-embed the input and output microstrip feeding lines effects and eliminate any impedance mismatch.

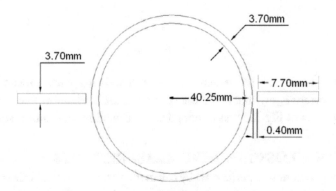

Figure 2.3: Microstrip ring resonator configuration diagram.

Fig. 2.3 shows a layout of the ring resonator along with the dimensions for the microstrip feeding lines, the gap in between the microstrip lines and the microstrip ring resonator, the width of the signal lines, and the mean radius r_m. Fig. 2.4 shows fabricated ring resonators with the TRL lines. S_{21} magnitude vs. frequency data were then inserted in a Mathcad program and the dielectric constant and loss tangent were extracted [4, 8]. A plot of S_{21} vs. frequency is shown in Fig. 2.5.

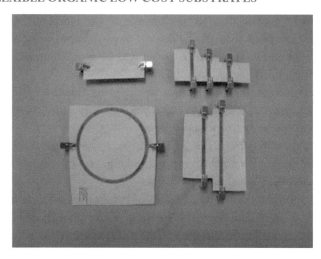

Figure 2.4: Photo of fabricated Microstrip ring resonators and TRL lines bonded to SMA connectors.

2.2.1   DIELECTRIC CONSTANT MEASUREMENTS

In order to extract the dielectric constant, the desired resonant peaks were first obtained according to [2, 8]:

$$f_o = \frac{nc}{2\pi r_m \sqrt{\varepsilon_{\text{eff}}}} \tag{2.1}$$

where f_o corresponds to the nth resonance frequency of the ring with a mean radius of r_m and effective dielectric constant ε_{eff} with c being the speed of light in vacuum. The extracted ε_r value at 0.71 GHz and 1.44 GHz of Fig. 2.5 was obtained using Equation (2.1) and is shown in Table 2.1.

2.2.2   DIELECTRIC LOSS TANGENT MEASUREMENTS

The extraction of loss tangent was performed by calculating the theoretical values of conductor and radiation losses. This is done in order to isolate the dielectric loss α_d since the ring resonator method gives the total loss at the frequency locations of the resonant peaks. The loss tangent is a function of α_d (in Nepers/m) according to [9]:

$$\tan \delta = \frac{\alpha_d \alpha_o \sqrt{\varepsilon_{\text{eff}} (\varepsilon_r - 1)}}{\pi \varepsilon_r (\varepsilon_{\text{eff}} - 1)} \tag{2.2}$$

where λ_o is the free-space wavelength, ε_r and ε_{eff} are the same as described above.

Available theoretical methods for calculating conductor loss and radiation loss have been dated from the 1970s [9]. $\tan \delta$ results are shown in Table 2.1 after subtracting the calculated conductor and radiation losses.

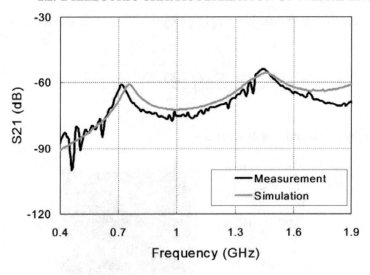

Figure 2.5: S_{21} vs. frequency for the ring resonator.

It is to be noted that the density of the paper substrate slightly increases after the bonding process described above [2]. This may slightly increase the calculated dielectric properties in Table 2.1 for multilayer paper-based RF modules.

Table 2.1: Extraction of dielectric constant from Ring Resonator Measurement

Mode	Resonant Freq (f_o)	Insertion Loss ($|S_{21}|$)	BW_{3dB}	ε_r	$\tan \delta$
n=1	0.71 GHz	−61.03 dB	42.12 MHz	3.28	0.061
n=2	1.44 GHz	−53.92 dB	75.47 MHz	3.20	0.053

2.2.3   CAVITY RESONATOR METHOD

When frequency range extends to 30 GHz, the roughness of the metal surface potentially approaches the skin depth, resulting in an inaccurate loss tangent extraction which usually requires acceptable theoretical equations for microstrip conductor losses [9]. In this case, the cavity resonator method provides a higher level of accuracy compared with the other methods, and has no requirement of a pretreatment on the substrate.

A split-cylinder resonator was fabricated with a circular-cylindrical cavity of radius 6.58 mm and length 7.06 mm, separated into two halves by a variable gap height which is adjustable to the thickness of the paper substrate being characterized, as shown in Fig. 2.6. The feeding structure is composed of coaxial cables terminated in coupling loops. A TE_{011} resonant mode was excited in the cavity at

$$f_{011} = \frac{3 \times 10^8}{2\pi} \sqrt{\left(\frac{3.8317}{a}\right)^2 + \left(\frac{\pi}{L}\right)^2}$$

(2.3)

where a is the cavity radius and L is its length.

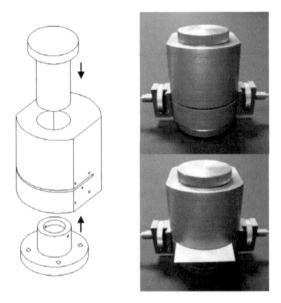

Figure 2.6: Cavity resonator in unloaded and loaded status.

A single sheet of the same hydrophobic paper was placed in the gap between the two cylindrical-cavity sections. The perturbation due to the inserted substrate caused the shifting of the TE_{011} resonant mode. Using the resonance and boundary conditions for the electric and magnetic fields, the substrate's dielectric constant can be calculated from the shifting [11]. The full wave electromagnetic solver HFSS was used to assist identifying the correct position of the TE_{011} resonant peak, as shown in Fig. 2.7. The measurement data of the resonant modes' shifting is plotted in Fig. 2.8. For the empty cavity, the dominant mode TE_{011} was observed at 34.54 GHz. After the paper sample was inserted, the TE_{011} shifted down to 33.78 GHz. In this way, the sample dielectric constant of ε_r = 1.6 was determined. Therefore, the relative permittivity of paper decreases with increasing frequency.

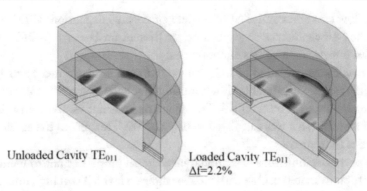

Figure 2.7: The simulated field distributions to help identifying the correct resonant peak corresponding to TE_{011} mode.

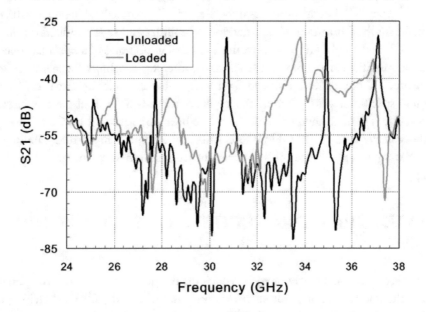

Figure 2.8: Measured modes shifting of the unloaded/loaded split-cylinder cavity.

2.3   LIQUID CRYSTAL POLYMER: PROPERTIES AND BENE-FITS FOR RF APPLICATIONS

Liquid Crystal Polymer (LCP) possesses attractive qualities as a high performance low-cost substrate and as a packaging material for numerous applications such as RFID/WSN modules, antenna arrays, microwave filters, high Q-inductors, RF MEMs and other applications extending throughout the mm-wave frequency spectrum. Furthermore, LCP has low loss, flexible, near hermetic nature, thermal stability, low cost and controlled coefficient of thermal expansion (CTE) in x-y direction make it one of the best candidates as a substrate for System on Package (SOP) approach for 3D integrated RF and mm-wave functions and modules.

The dielectric characterization of LCP substrate has been performed up to 110 GHz using several methods that are regarded as highly accurate and include: ring resonator, cavity resonator, as well as a transmission line TL method [9]. The dielectric constant vs. frequency show a value for ε_r = 3.16 ± 0.05 and the tan δ was calculated to be <0.0049. This proves the excellence of LCP in electrical properties for mm-waves.

LCP also possesses numerous exceptional mechanical properties not to mention that it's an environmentally friendly material. One particular example and which is of high interest in mm-wave is the CTE. LCP can be engineered to have an x-y CTE between 0 ppm/°C and 40 ppm/°C and so this unique process can achieve a thermal expansion match in the x-y plane with many commonly used material such as Cu (16.8 ppm/°C), Au (14.3 ppm/°C), Si (4.2 ppm/°C), GaAs (5.8 ppm/°C), and SiGe (3.4-5 ppm/°C). On the other hand the z-axis CTE is considerably high (~105 ppm/°C); however and due to the fabrication miniaturization capabilities and LCP productions thin layers of LCP (2 mils LCP and 1 mil LCP bond ply), the z-expansion becomes of a minimal concern unless thick multilayer modules come into consideration. Another key capability for many applications in RF is the assumed ability of flexibility and light weight, for examples antennas on LCP may be conformed into specific shapes desired by the application needs. LCP is also a near hermetic, low water permeability. Thermal stability results have also been obtained and LCP has been proven to be as good as or better than the 10 GHz PTFE/glass and alumina temperature stability values [9]. This verifies the superiority of using this material for RF and mm-wave integrated modules, SOP, or packaging.

2.4   INKJET-PRINTING TECHNOLOGY AND CONDUCTIVE INK

Inkjet printing is a direct-write technology by which the design pattern is transferred directly to the substrate, and there is not requirement of masks compared with the traditional etching technique which has been widely used in industry. Besides that, unlike etching which is a subtractive method by removing unwanted metal from the substrate surface, inkjet printing jets a single ink droplet from the nozzle to the desired position, therefore, no waste is created, resulting in an economical fabrication solution.

Silver nano-particle inks are usually selected in the inkjet-printing process to ensure a good metal conductivity. Silver ink is sprayed from the cartridge nozzles to the substrate. The operation of the jetting system, illustrated in Fig. 2.9, is based on voltage applied at the orifice of each nozzle. The spraying of silver ink droplets is controlled by the automatic adjusting of the voltage in the charge electrode and across the deflection plates. When the nozzles are not jetting, a voltage is still applied, so that the ink is contained at the edge of the nozzles and is not dripping down to the substrate.

Manual setting of the nozzle voltage can be applied in order to control the thrust and speed of the ink drops.

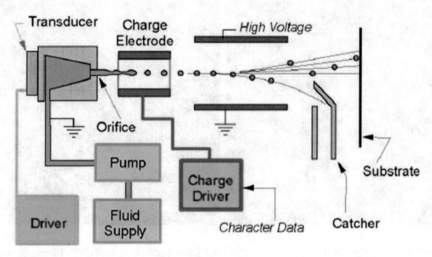

Figure 2.9: Inkjet-printing mechanism.

After the silver nano-particle droplet is driven through the nozzle, sintering process is found to be necessary to remove excess solvent and to remove material impurities from the depositions. Sintering process also provides the secondary benefit of increasing the bond of the deposition with the paper substrate [12]. The conductivity of the conductive ink varies from $0.4 \sim 2.5 \times 10^7$ Siemens/m depending on the curing temperature and duration time [13]. Fig. 2.10 shows the difference between heating temperature 100°C and 150°C after a 15 minutes curing. At lower temperature, large gap exists between the particles, resulting in a poor connection. When the temperature is increased, the particles begin to expand and gaps start to diminish. This guarantees a virtually continuous metal conductor, providing a good percolation channel for the conduction electrons to flow. The silver nano-particle ink electrical performance versus cure time at temperature in air is shown in Fig. 2.11.

There is also a difference between sintering a thin film layer and a bulk form. The temperature distribution can be assumed to be a constant in a thin film layer, however, a significant temperature gradient in the bulk form, is resulting in a different conductivity distribution inside the inkjet-printed layers. Bulk inkjet-printed layer which allows the realization of the right metal thickness is the form used to ensure the conductivity performance of microwave circuits, such as RFID module and multilayer bandpass filters. Curing temperature of 120°C and duration time of two hours is recommended in the following fabrications to sufficiently cure the nano-particle ink.

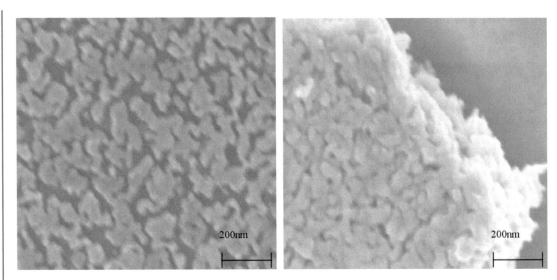

Figure 2.10: SEM images of a layer of printed silver nano-particle ink, after a 15 minutes curing at 100°C and 150°C, respectively. At higher temperature, gaps between nano-particles diminish, forming a continuous metal layer for the electrons to flow.

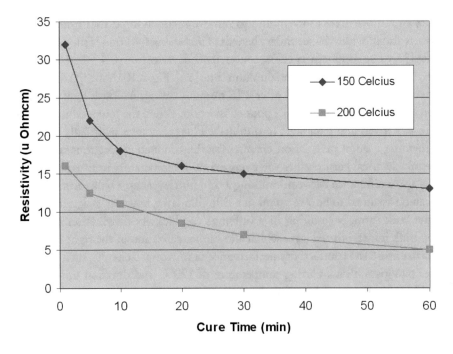

Figure 2.11: Silver nano-particle ink electrical performance versus cure time [13].

Bibliography

[1] M. Lessard, L. Nifterik, M. Masse, J. Penneau, and R. Grob, R, *"Thermal aging study of insulating papers used in power transformers,"* Electrical Insulation and Dielectric Phenomena 1996, IEEE Annual Report of the Conference on, Vol. 2, pp. 854–859, 1996. DOI: 10.1109/CEIDP.1996.564642

[2] L. Yang, A. Rida, R. Vyas, and M. M. Tentzeris, *"RFID Tag and RF Structures on a Paper Substrate Using Inkjet-Printing Technology,"* Microwave Theory and Techniques, IEEE Transactions on Volume 55, Issue 12, Part 2, pp. 2894–2901, Dec. 2007. DOI: 10.1109/TMTT.2007.909886

[3] A. Rida, L. Yang, R. Vyas, S. Basat, S. Bhattacharya, and M. M. Tentzeris, *"Novel Manufacturing Processes for Ultra-Low-Cost Paper-Based RFID Tags With Enhanced "Wireless Intelligence""* Proc. Of the 57th IEEE-ECTC Symposium, pp. 773–776, Sparks, NV, June 2007. DOI: 10.1109/ECTC.2007.373885

[4] L. Yang, and M. M. Tentzeris, *"3D Multilayer Integration and Packaging on Organic/Paper Low-cost Substrates for RF and Wireless Applications"* ISSSE '07. International Symposium on Signals, Systems and Electronics, 2007. July 30 2007-Aug. 2, pp. 267–270, 2007. DOI: 10.1109/ISSSE.2007.4294464

[5] M. M. Tentzeris, L. Yang, A. Rida, A. Traille, R. Vyas, and T. Wu, *"RFID's on Paper using Inkjet-Printing Technology: Is it the first step for UHF Ubiquitous "Cognitive Intelligence" and "Global Tracking"?"* RFID Eurasia, 2007 1st Annual 5-6, pp. 1–4, Sept. 2007. DOI: 10.1109/RFIDEURASIA.2007.4368098

[6] A. Rida, R. Vyas, S. Basat, A. Ferrer-Vidal, L. Yang; S. Bhattacharya, and M. M. Tentzeris, *"Paper-Based Ultra-Low-Cost Integrated RFID Tags for Sensing and Tracking Applications"* Electronic Components and Technology Conference, 2007. 57th, pp. 1977–1980, May 29 2007-June 1 2007. DOI: 10.1109/ECTC.2007.374072

[7] M. M. Tentzeris, L. Yang, A. Rida, A. Traille, R. Vyas, and T. Wu, *"Inkjet-Printed RFID Tags on Paper-based Substrates for UHF "Cognitive Intelligence" Applications"* IEEE International Symposium on Personal, Indoor and Mobile Radio Communications, pp. 1–4, 3-7 Sept. 2007. DOI: 10.1109/PIMRC.2007.4394346

[8] A. Rida, L Yang; R. Vyas, S. Bhattacharya, and M. M. Tentzeris, "*Design and integration of inkjet-printed paper-based UHF components for RFID and ubiquitous sensing applications*" IEEE Microwave European Conference, pp. 724–727, Oct. 2007. DOI: 10.1109/EUMC.2007.4405294

[9] D. Thompson, O. Tantot, H. Jallageas, G. Ponchak, M. M. Tentzeris, and J. Papapolymerou, "*Characterization of LCP material and transmission lines on LCP substrates from 30 to 110 GHz*," IEEE Trans. Microwave Theory and Tech., vol. 52, no. 4, pp. 1343–1352, April 2004. DOI: 10.1109/TMTT.2004.825738

[10] S. Basat, S. Bhattacharya, A. Rida, S. Johnston, L. Yang, M. M. Tentzeris, and J. Laskar, "*Fabrication and Assembly of a Novel High-Efficience UHF RFID Tag on Flexible LCP Substrate*," Proc. of the 56th IEEE-ECTC Symposium, pp. 1352–1355, May 2006. DOI: 10.1109/ECTC.2006.164583

[11] G. Kent, "An evanescent-mode tester for ceramic dielectric substrates," IEEE Trans. Microwave Theory & Tech., vol. 36, no. 10, pp. 1451–1454, October 1998. DOI: 10.1109/22.6095

[12] M. Carter, J. Colvin, and J. Sears, "Characterization of conductive inks deposited with maskless mesoscale material deposition," TMS2006, Mar. 12-16, San Antonio, Texas, USA.

[13] Cabot Corporation, "Inkjet silver conductor AG-IJ-G-100-S1," Data Sheet, October 2006.

CHAPTER 3

Benchmarking RFID Prototypes on Organic Substrates

3.1 RFID ANTENNA DESIGN CHALLENGES

A major challenge in RFID antenna designs is the impedance matching of the antenna (Z_{ANT}) to that of the IC (Z_{IC}). For years, antennas have been designed primarily to match either 50 Ω or 75 Ω loads. However, RFID chips primarily exhibit complex input impedance, making matching extremely challenging [1].

It is to be noted that besides impedance matching, low cost, omnidirectional radiation pattern, long read range, wide bandwidth, flexibility, and miniaturized size are all important features that an RFID tag must acquire. Most available commercial RFID tags are passive due to cost and fabrication requirements. A purely passive RFID system utilizes the EM power transmitted by the reader antenna in order to power up the IC of the RFID tag and transmit back its information to the reader using the backscatter phenomena. A block diagram of a passive RFID tag is shown in the Fig. 3.1 below. The antenna matching network must provide the maximum power delivered to the IC which is used to store the data that is transmitted to/received from the reader.

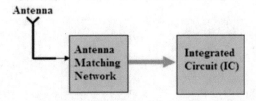

Figure 3.1: Block diagram of a passive RFID tag.

For a truly global operation of passive UHF RFID's, Gen2 protocols define different sets of frequency, power levels, numbers of channel and sideband spurious limits of the RFID readers signal, for different regions of operation (North America 902-928 MHz, Europe 866-868 MHz, Japan 950-956 MHz, and China 840.25-844.75 MHz and 920.85-924.75 MHz). This places a demand for the design of RFID tags that operate at all those frequencies, thus requiring a miniaturized broadband UHF antenna. For instance, in a scenario where cargo/containers get imported/exported from different regions of the world in a secured RFID system implementation, an RFID tag is required to have a bandwidth wide enough to operate globally. This imposes very stringent design challenges on the antenna designers.

3.2   RFID ANTENNA WITH SERIAL STUB FEEDING STRUCTURES

3.2.1   DESIGN APPROACH

In this design approach, a compact configuration (in an area less than 3" x 3") of a $\lambda/2$ dipole antenna was developed, where λ is the free space wavelength. The antenna design is favorable for its quasi-omnidirectional radiation pattern. The step by step design is illustrated in the figures below. Fig. 3.2 shows the main radiating element which if stretched from one end to the other corresponds to a length of ~16cm (which is $\lambda/2$ around the center frequency 935 MHz in air). The tapering of the antenna was chosen for maximum current flow (hence optimum efficiency) and to achieve a high bandwidth. The $\lambda/2$ antenna was folded as shown in Fig. 3.2 at a distance (~0.16 λ) not to cause any significant current perturbation, while making the design more compact. Without loss of generality, in this design the overall matching network is designed to conjugately match an RFID chip with a high capacitive impedance of $Z_{IC} = 73 - j113\,\Omega$. Fig. 3.2 also shows the step by step procedure used in the design. To satisfy the conformality RFID requirements, the proposed antenna was fabricated on flexible 4-mil LCP. The IC used for this design has four ports; two input ports namely RF1 and RF2 are identical and may be connected to a single or dual antenna configuration. The resistive shorting stub and the double inductive stub as illustrated in Fig. 3.3 constitute the overall matching network. The resistive stub is used to tune the resistance of the antenna to match that of the IC. In this design the size and shape (thin long loop shaped line) of the resistive stub were designed to have an optimum match to $Z_{IC} = 73 - j113\,\Omega$. The double inductive stub is composed of two inductive stubs to provide symmetry on both sides of the antenna. The double inductive stub also serves as the reactive tuning element of the antenna.

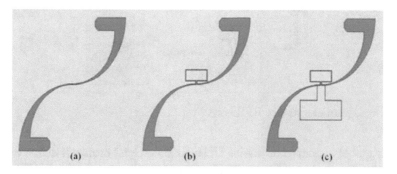

Figure 3.2: Step by step antenna design showing (a) Radiating body (b) Radiating body plus double inductive stub and (c) final antenna structure with the resistive stub.

The feeding point of the antenna is at the bottom part of the double inductive stub where an IC would be surface mounted. Fig. 3.3 illustrates the final structure.

The stubs were designed to have a center frequency f_0 at 895 MHz with a bandwidth of 70 MHz operating from 860→930 MHz (European and U.S. frequencies). A wide frequency sweep has also been performed up to 5 GHz where no parasitic radiation has been observed for this antenna. Those variables can be fine tuned to optimize the antenna characteristics on the RFID tag at any frequency and matched to any IC impedance.

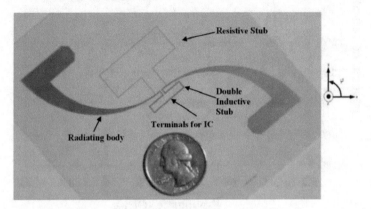

Figure 3.3: RFID antenna structure showing stubs.

The structure was simulated and optimized in the system level design tool HFSS. The input impedance of the simulated antenna design is shown in Fig. 3.4. As it can be observed the RFID UHF band (860→930 MHz) is outside the antenna self-resonance peak, resulting in a more flat impedance response against frequency. This yields to a bandwidth of ∼8% which is predominantly realized by the finite slope of the reactance of the antenna in the frequency of interest.

The simulated impedance at the center frequency $f_0 = 895$ MHz is $57.46 + j112.1\ \Omega$ which results in a return loss RL< −18 dB. This antenna has a bandwidth of ∼8% (70 MHz) where the bandwidth is defined by a Voltage Standing Wave Ratio (VSWR) of 2 (alternatively a RL of −9.6 dB) as shown in the Results and Discussion section.

The return loss of this antenna was calculated based on the power reflection coefficient which takes into account the capacitance of the IC [2]:

$$\left| s^2 \right| = \left| \frac{Z_{IC} - Z_{ANT}{}^*}{Z_{IC} + Z_{ANT}} \right|^2 \tag{3.1}$$

Z_{IC} represents the impedance of the IC and Z_{ANT} represents the impedance of the antenna with $Z_{ANT}{}^*$ being its conjugate.

The simulated radiation pattern and radiation efficiency were numerically computed in HFSS by introducing an RLC boundary along with the port impedance that simulates the behavior of the IC (with its complex impedance feed). In Fig. 3.5 the 2-D radiation plot is shown for the phi=0° and phi=90° where an omnidirectional pattern is realized. The radiation pattern throughout the bandwidth of the antenna has also shown to have an omnidirectional pattern similar to that of a classic ($\lambda/2$) dipole antenna.

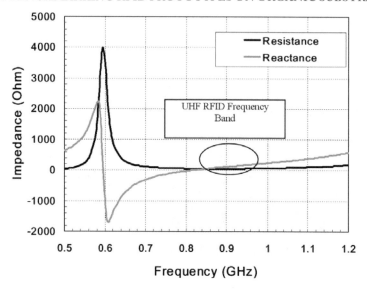

Figure 3.4: Simulated input impedance of the S-shaped antenna.

A directivity of 2.10 dBi is achieved with a radiation efficiency of 97%. The omnidirectional radiation is one of the most fundamental requirements for RFID's to allow for their reading/writing operation independent of the orientation of their antenna with respect to the reader.

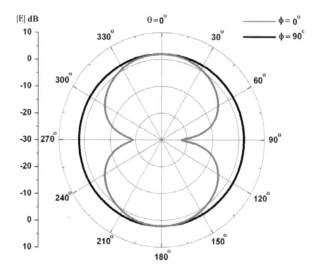

Figure 3.5: 2D far-field radiation plot for S-shape antenna.

3.2.2   ANTENNA CIRCUIT MODELING

In order to obtain a thorough understanding of the power reflection caused by any mismatch at the terminals of the feed structure of the antenna, a wideband equivalent circuit model has been developed. This model serves as a benchmark for the design of an RFID antenna to theoretically match any Z_{IC} for maximum power flow resulting in optimum antenna efficiency and an excellent read range.

Based on a physical approach, an equivalent lumped element circuit model was derived. The system level design and simulation tool Advanced Design System (ADS) was used to simulate the behavior of the circuit model (S_{11} parameter) and resulted in a negligible error function ($< 10^{-5}$). Fig. 3.6 below shows the agreement between the lumped element circuit S parameters with that of the structure (from the full wave simulator).

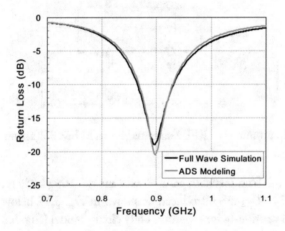

Figure 3.6: S_{11} for the exact structure and for the equivalent circuit model.

Fig. 3.7 shows the detailed equivalent circuit of the radiating body only. Each arm of the radiating body consists of a resistor in series with an inductor, the combination of which models the metal effects. A capacitor in series with a resistor, which are located in parallel with the previous combination of L_s and R_s model the substrate effects. Finally, the capacitive coupling or E-Field coupling between the two arms of the S-shaped antenna is modeled by the top and bottom capacitors (for air and dielectric capacitive coupling, respectively). This lumped element model covers a frequency range 700$\rightarrow$1100 MHz as shown in Fig. 3.6 above.

The circuit configuration in Fig. 3.7 can be simplified to the one shown in Fig. 3.8 by using symmetry and direct circuit analysis.

Since the double inductive stub is connected in series with the radiating body, the equivalent circuit model of the second stage design (radiating body plus inductive stub) as shown in Fig. 3.2 has the same circuit elements configuration as the one in Fig. 3.8, with change in values only. The final stage of the design has the circuit model configuration shown in Fig. 3.9. Due to the configuration

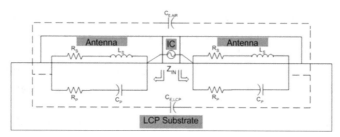

Figure 3.7: Cross-sectional detail showing equivalent lumped element model of RFID antenna shown in Fig. 3.2(a) and 3.2(b).

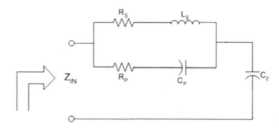

Figure 3.8: Equivalent circuit model of RFID antenna shown in Fig. 3.2(a) and 3.2(b).

of the resistive shorting stub (connected in parallel with the radiating body plus inductive stub), the components: R_{s2}, L_{s2}, R_{p2}, C_{p2} are introduced as shown in Fig. 3.17 below and model the same effects as those discussed previously for radiating body circuit model (Fig. 3.8).

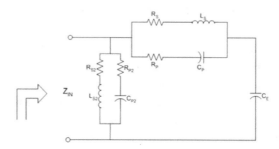

Figure 3.9: Equivalent circuit for antenna structure shown in Fig. 3.2(c).

The equivalent circuit shows how stubs can be used to tune the impedance in order to match to any IC. Parametric sweeps can be used along different stubs structures (for example loops structures can be used for adding series inductance or parallel capacitance). The resistance of the antenna is mainly determined by the radiating body and can be tuned by the two stubs as shown above. This model also helps to determine the amount of loss (as parallel resistance and capacitance) due to the substrate loss which helps in understanding radiation efficiency as a function of the substrate.

3.2.3   MEASUREMENT RESULTS AND DISCUSSION

In order to accurately measure the input impedance of the RFID antennas, numerous problems should be taken into consideration. First of all, a traditional probe station was not suitable for our tests due to the undesired shorting effect of the metallic chuck, which was behaving as a spurious ground plane for the dipole antennas. To tackle the problem, a custom-made probe station using wood and high density polystyrene foam was built. This type of foam was selected due to its low ε_r of 1.06 [3] resembling that of the free space. A $5 \cdot \lambda/2$-thick foam station was designed in order to ensure minimum backside reflections of the antenna.

It was also taken into account the fact that the antennas were balanced structures and a typical GS probe connected to a regular coaxial cable would provide an unbalanced signal, as shown in Fig. 3.10. To prevent a current difference between the dipole arms, a $\lambda/4$ balun with an operational bandwidth of $840 \rightarrow 930$ MHz (which covers the band of interest for the design) was used. After all the above mentioned precautions were taken and minding about the calibration process, S-parameters were measured using Agilent 8510C VNA and transformed to Z_{IN} or Z_{ANT}.

Figs. 3.11 and 3.12 show a very good agreement between the simulated results and the measurements for the antenna input impedance and S_{11} parameter, respectively. The demonstrated antenna bandwidth allows for a universal operation of the proposed UHF RFID (Worldwide frequency coverage except Japan and some Asian countries that operate at a frequency of 950 MHz and higher) frequency band.

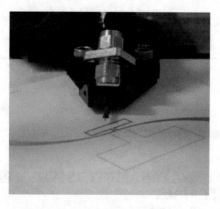

Figure 3.10: Photograph of the probe plus S-shaped antenna.

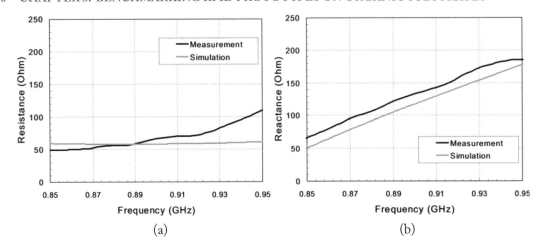

Figure 3.11: Measured and simulated data of input impedance: (a) Resistance (b) Reactance.

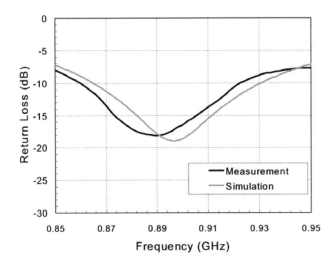

Figure 3.12: Measured and simulated data of return loss.

3.2.4   EFFECT ON ANTENNA PARAMETERS WHEN PLACED ON COMMON PACKAGING MATERIALS

In order to fully investigate the effect of the surroundings on the antenna parameters, such as the resonance, bandwidth, and radiation; the S-shaped antenna was simulated for 3 practical configurations: on a 4 mm thick common plastic material [PET-Polyethylene terephthalate (ε_r=2.25 and tan $\delta = 0.001$)], on 4 mm thick paper (ε_r=3.28 and tan $\delta = 0.006$ [4]) substrate as well in an embedded structure with 0.5 mm thick paper superstrate on a 4 mm paper substrate. Fig. 3.13 shows the Return Loss results for the polyethylene and paper substrates. A shift in resonance frequency occurs (95 MHz for paper and 60 MHz for PET from the original antenna with center frequency 895 MHz on LCP substrate). This observation can be easily corrected by scaling down the x-y dimensions of the antenna. In the paper case the antenna was scaled down by 13% while the antenna on PET by a factor of 8% and the new Return Loss results as shown in Fig. 3.14 were obtained. As seen in the figure, the detuning of the resonance can be easily performed by scaling the whole structure. For example, when placed on paper substrate, detuning becomes necessary if the thickness exceeds 1.5 mm.

As for the most common case where these RFID tags are placed on cardboard boxes; the dielectric properties of cardboard do not impede antenna characteristics due to its low dielectric constant (close to 1 and low loss properties [5]). However, the effect of the enclosed materials and the distance of the RFID tag from the arbitrarily placed enclosed contents play a more important role than the size and thickness of the cardboard. An alternative way to increase the bandwidth of the antenna in order to compensate for the material/fabrication variations is the use of a more broadband matching section, potentially introducing an additional stub-line.

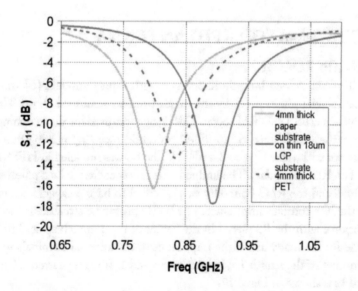

Figure 3.13: Return loss for 4 mm thick paper and PET substrates.

In order to analyze the effect of the radiation pattern of these materials, the gain for the S-shaped antenna has been plotted in Fig. 3.15 for the LCP and paper substrates. The worst case scenario observed for gain loss was 1.049 dB (on 4 mm paper) for the E-plane or $\varphi=0$ degrees plane in comparison with 2.095 dB (on 18 μm LCP).

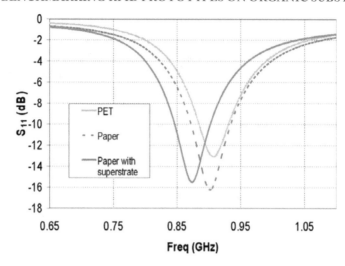

Figure 3.14: Return loss for 4 mm thick paper and PET substrates with modified Antenna Dimensions.

3.3   BOWTIE T-MATCH RFID ANTENNA

3.3.1   DESIGN APPROACH

In this section a T-match folded bow-tie half-wavelength dipole antenna [6] was designed and fabricated on a commercial photo paper by the inkjet-printer mentioned above. The antenna was designed using Ansoft's HFSS 3-D EM solver. This design was used for the matching of the passive antenna terminals to the TI RI-UHF-Strap-08 IC with resistance $R_{IC} = 380$ Ohms and reactance modeled by a capacitor with value $C_{IC} = 2.8$ pF [3]. The IC was modeled in HFSS by introducing a lumped port and an RLC boundary. The lumped port was specified to be a purely resistive source with $R = 380$ Ohms and the RLC boundary was specified to have a capacitance value of 2.8 pF; hence simulating the IC's complex impedance. The RFID prototype structure is shown in Fig. 3.16 along with dimensions, with the IC placed in the center of the T-match arms. The T-match arms are also responsible for the matching of the impedance of the antenna terminals to that of the IC through the fine tuning of the length L_3, height h, and width W_3. The current distribution of this antenna at 900 MHz is shown in Fig. 3.17.

3.3.2   RESULTS AND DISCUSSION

A GS 1000 μm pitch probe connected to a UHF balun to ensure the balanced signal between the arms of the T-match folded dipole antenna was used for impedance measurements, as shown in Fig. 3.18. In order to minimize backside reflections of this type of antenna, the fabricated or inkjet-printed antennas were placed on a custom-made probe station as discussed in Section 3.2.3. The calibration method used was short-open-load-thru (SOLT). Fig. 3.19 shows the impedance plots.

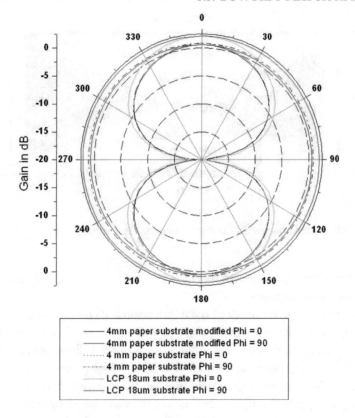

Figure 3.15: Radiation Pattern of the Gain of the s-shaped antenna on paper, and LCP substrates.

As shown in Fig. 3.19(a), the simulated resistance for the antenna in the UHF RFID frequency range maintains a value close to 380 Ohms between the two successive peaks. The reactance part of the impedance, as shown in Fig. 3.19(b), features a positive value with a linear variation with frequency, pertaining to an inductance that conjugately matches or equivalently cancels the effect of the 2.8pF capacitance of the IC. Fairly good agreement was found between the simulation and measurement results. The distortion is possibly due to the effect of the metal probe fixture.

The return loss of this antenna was calculated based on the power reflection coefficient as shown in Equation (3.1) in Section 3.2.1. The plot is shown in Fig. 3.20 demonstrating a good agreement for both paper metallization approaches. The nature of the bow-tie shape of the half-wavelength dipole antenna body allows for a broadband operation, with a designed bandwidth of 190 MHz corresponding to 22% around the center frequency 854 MHz which covers the universal UHF RFID bands. It has to be noted that the impedance value of the IC stated above was provided only for the UHF RFID frequency which extends from 850 MHz to 960 MHz; thus, the return

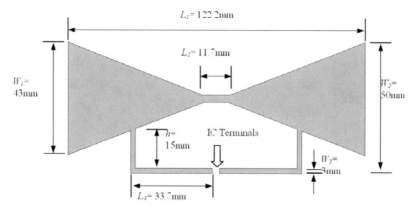

Figure 3.16: T-match folded bow-tie RFID tag module configuration.

loss outside this frequency region, shown in Fig. 3.20, may vary significantly due to potential IC impedance variations with frequency.

In order to verify the performance of the inkjet-printed RFID antenna, measurements were performed on a copper-metalized antenna prototype with the same dimensions fabricated on the same paper substrate using the slow etching technique mentioned before. The return loss results are included in Fig. 3.20 and they show that the return loss of the inkjet-printed antenna is slightly larger than the copper one. Overall a good agreement between the copper etched and the inkjet-printed antennas was observed despite the higher metal loss of the silver-based conductive ink.

The radiation pattern was measured using Satimo's Stargate 64 Antenna Chamber measurement system as shown in Fig. 3.21 and by using the NIST Calibrated SH8000 Horn Antenna as a calibration kit for the measured radiation pattern at 915 MHz. The radiation pattern as shown in Fig. 3.22 is almost uniform (omnidirectional) at 915 MHz with directivity around 2.1 dBi. The IC strap was attached to the IC terminal with H2OE Epo-Tek silver conductive epoxy cured at 80°C. An UHF RFID reader was used to detect the reading distance at different directions to the tag. These measured distances are theoretically proportional to the actual radiation pattern. The normalized radiation patterns of simulation, microwave chamber measurement and reader measurement are plotted in Fig. 3.22, showing a very good agreement between simulations and measurements, which can be also verified for other frequencies within the antenna bandwidth.

3.4   MONOPOLE ANTENNA

Minor drawbacks that might occur with the dipole based module can be eliminated by using a monopole based structure. Unlike the dipole antenna which has no ground, the monopole uses its ground planes as a radiating surface, which can also be used to shield any circuitry behind it. The monopole antenna also does not require a differentially fed input signal like the dipole, which is ideal

Figure 3.17: Simulation plot of current distribution at 900 MHz.

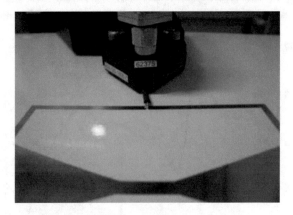

Figure 3.18: Photograph of the impedance measurement using GS pitch probe.

for Power Amplifiers (PA) since their output might be single-ended. This section gives guidelines on designing a monopole antenna that can be used for Identification and Sensing applications.

3.4.1   DESIGN APPROACH

The geometry and a photograph of the proposed CPW-fed printed monopole antenna is shown in Fig. 3.23. This module uses paper as a substrate, with a thickness of 0.254 mm, and overall dimensions 75 mm (width) x 100 mm (length), including the feeding line. A Coplanar Waveguide (CPW) transmission line consisting of a single metallic layer is selected for feeding the antenna because of its easy integration on the paper substrate due to its planar structure. The antenna structure is composed of a planar Z-shaped rectangular monopole with width 50 mm, length 56 mm and a spacing of $h = 11$ mm from the ground plane. Two rectangular slots are embedded into the radiating element from both side edges, resulting in a meander-like antenna as shown in Fig. 3.23. Both slots have a width of 10 mm and lengths of $l_1 = l_2 = 40$ mm, chosen to optimize the matching of the antenna to the load.

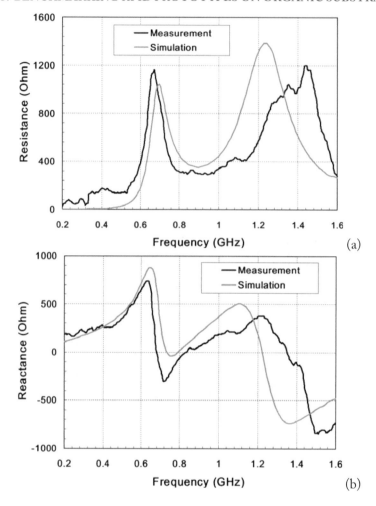

Figure 3.19: Measured and simulated input resistance and reactance of the inkjet-printed RFID tag, (a) Resistance (b) Reactance.

As discussed earlier, RFID antenna design should fulfill several design requirements: "global-operability" UHF bandwidth [USA (902–928 MHz), Europe (865–868 MHz)] RFID band, omni-directional radiation pattern, impedance matching with RFID circuitry, long read range, and compact size. Optimizing these parameters involves however inevitable tradeoffs between them [7]. Maximizing the size of the ground plane increases the directivity of the antenna, because the ground acts as a radiating element and also shields it from the rest of electronic circuitry; however it increases the profile of the antenna. Impedance matching in RFID tags between the antenna and the load is very significant for the range maximization. In order to ensure maximum power transfer from the antenna to the load, the antenna was matched to an impedance of 60.1-j73.51 ohms (Z_{L-opt})

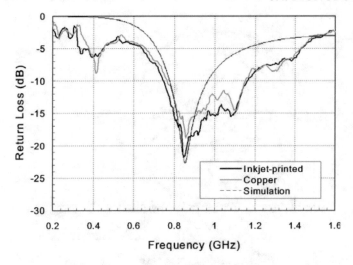

Figure 3.20: Measured and simulated Return Loss of the inkjet-printed RFID tag.

Figure 3.21: Photograph of the radiation pattern measurement in an Antenna Chamber.

which is the reference at the PA output (to be integrated with this antenna module) at the design frequency of 904.4 MHz. The parameters h, g, and w were also optimized to fine tune the desired antenna impedance and increase the directivity. The height (h) of the radiating element from the ground has a major influence on the performance of the antenna, as it modifies the radiation pattern and the impedance of the antenna. By increasing it, guided waves of the antenna are transitioning more efficiently into free-space waves and the impedance becomes more capacitive. Maximizing it however results into thick and difficult to mount RFID tags. The value of the parameters after optimization where found $h = 11$ mm, $g = 0.3$ mm and $w = 3.8$ mm.

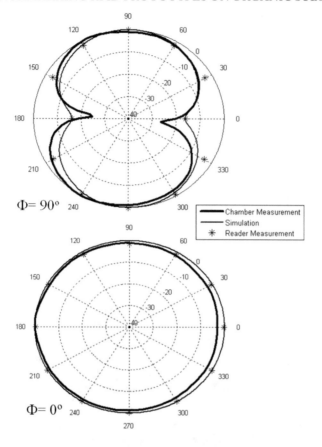

Figure 3.22: Normalized 2D far-field radiation plots of simulation, chamber measurement and tag reading distance measurement.

3.4.2   RESULTS AND DISCUSSION

The performance of two prototypes (inkjet-printed and copper tape fabricated) was experimentally tested and the results are presented in Fig. 3.24, which shows the simulated and measured Return Loss featuring a good agreement. It can be seen from the simulations that the inkjet-printed antenna has a resonance at 904 MHz and a −10 dB impedance bandwidth of 132 MHz (822–954 MHz) corresponding to 14.6% around the resonant frequency. The measured S_{11} plot of the inkjet-printed antenna is resonant at 898 MHz with bandwidth of 82 MHz (860–942 MHz) corresponding to 9.1% around the resonant frequency, where the copper tape fabricated antenna displays a return loss plot with resonant frequency at 906 MHz and a bandwidth of 68 MHz (874 – 942 MHz) corresponding to 7.5%. The radiation characteristics of the proposed antenna have also been investigated and are depicted in Fig. 3.25. The simulated radiation patterns for the x-z plane and the y-z plane at the resonant frequency 904.5 MHz demonstrate a radiation pattern similar to a conventional monopole

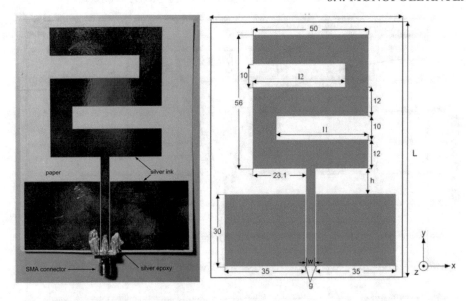

Figure 3.23: Photograph and Schematic (units in mm) of the proposed monopole antenna.

antenna, displaying an omnidirectional radiation pattern on the x-z plane and a directional pattern with 2 nulls on the y-z plane. The directivity of the antenna was found 0.2 dBi in simulation.

3.4.3   ANTENNA GAIN MEASUREMENT

The following equipment was used to carry out the gain measurements for the inkjet-printed and copper fabricated UHF monopole antenna:

ROHDE & SCHWARZ SMJ 100A Vector Signal Generator

Tektronix RSA3408A DC – 8 GHz Real Time Spectrum Analyzer

AN-400 Reader Antenna with gain G_r=6 dBi

Typical 3.5 mm Coaxial Cable with attenuation=0.1 dB

The equipment was set up as shown in Fig. 3.26. The prototype antenna to be measured was treated as the transmitter and was connected with a coaxial cable to the Vector Signal Generator. The antenna (transmitter) was attached at a height of $h_t = 1.60$ m on a piece of polystyrene foam using thin adhesive tape, as shown in Fig. 3.27. Polystyrene foam was used to hold the antenna in place, simulating a free-space environment. The AN-400 antenna is a commercial RFID reader antenna and is treated as the receiver in the experimental setup. The receiver was connected with a coaxial cable to the Real Time Spectrum Analyzer and was positioned at a fixed place that had also a height of $h_r = 1.60$ m, as shown in Fig. 3.28. Both distances were measured from the middle point

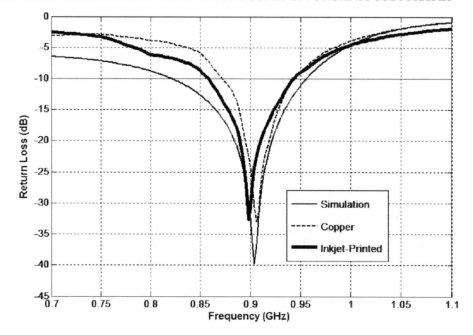

Figure 3.24: RL of the inkjet-printed and copper tape fabricated monopole antenna.

of each antenna to the ground. The distance between the transmitter and the receiver was measured $d = 4.06$ m.

Verifying that the receiver is at the far field region of the transmitter was performed as a first step using Fraunhofer's far field equation, shown below:

$$R = \frac{2D^2}{\lambda} \tag{3.2}$$

where:

R : the distance from the prototype antenna at which its far field region starts

D : the largest dimension of the antenna

λ : the free space wavelength at the transmission frequency of 904.5 MHz

The far field boundary for the monopole antenna is evaluated using Equation (3.2) to be $R = 6.0$ cm. Therefore, it is verified that the receiver is in the far field region of the monopole antenna. It is also essential to make sure that there is a clear line of sight between the two antennas and that no other objects are at a smaller distance from the antenna's far field boundary d_t. Finally, it is critical that both antennas are placed in the optimum orientation with each other. Each antenna must be placed in the direction of maximum radiation intensity or gain of the other. In this way the gain of the fabricated antenna can be measured correctly at the operating frequency and compared to the simulation. The gain measurement is performed for both the inkjet-printed and the copper

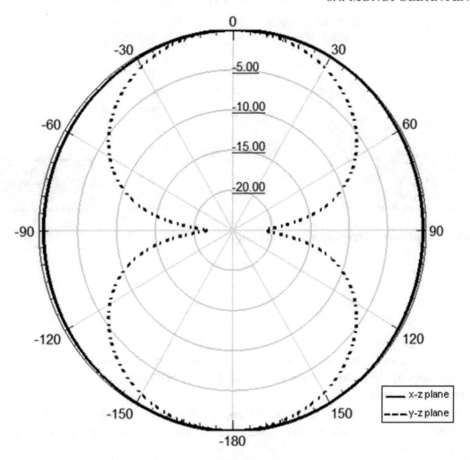

Figure 3.25: Simulated radiation pattern of monopole antenna at 904.5 MHz.

fabricated antenna. The Vector Signal Generator is set so that it transmits a continuous wave signal at the operating frequency of the antenna, which is at 904.5 MHz with power P_s =0.1 dBm. The Signal Generator is connected to the antenna, with a 50Ω coaxial cable with loss L_c =0.1 dB. Therefore, the power at the end of the coaxial cable and before the input of the antenna is $P_t = P_s - L_c = 0$ dBm.

In addition the antenna is not matched to the 50Ω coaxial cable, since it has complex input impedance. As a result there is another power loss in the antenna input because of power reflection, between the antenna and the 50Ω transmission line. The reflection coefficient ρ for the both inkjet-printed and copper fabricated antenna is calculated by Equation (3.3) taking into account their respective measured input impedance at the operating frequency.

$$\rho = \frac{Z_{ant} - Z_0}{Z_{ant} + Z_0} \qquad (3.3)$$

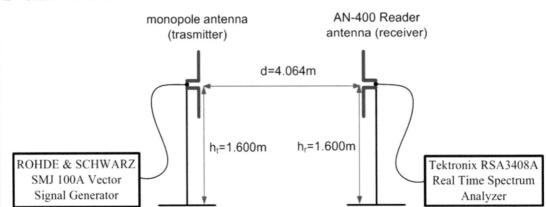

Figure 3.26: Gain measurement experimental setup.

where:

Z_{ant}: the measured input impedance of the antenna at the operating frequency

Z_0 : the characteristic impedance of the coaxial cable ($Z_0 = 50\,\Omega$)

ρ : reflection coefficient

The magnitude of ρ is found for the inkjet-printed antenna $|\rho| = 0.59$ and for the copper tape fabricated antenna $|\rho| = 0.60$. The mismatch loss is evaluated for both antennas as: $L_m = \left(1 - |\rho|^2\right)_{dB} = 1.9$ dB.

The input power after the losses is radiated from the proposed antenna (transmitter) into space and is received by the Reader Antenna (receiver). The reader antenna has gain of $G_r = 6$ dBi and is matched to the 50Ω coaxial cable that connects it to the spectrum analyzer. Because the receiver antenna is circular polarized and the transmitter monopole antenna is linear polarized, a polarization mismatch exists in the wireless link between the two antennas. The polarization mismatch has to be accounted and is assumed to be $L_p = 3$ dB. The accepted power from the reader antenna is measured by the spectrum analyzer, which was set to a center frequency of 904.5 MHz with a resolution bandwidth of 36 MHz. The measured accepted power results are the same for both antennas and are illustrated in Fig. 3.29.

In order to calculate the gain of the antenna from the measured results, a propagation model has to be applied. To determine which propagation model is suitable for our measurement setup, an investigation to whether any obstacles exist in the first Fresnel zone of our wireless link setup was performed. Therefore, the maximum radius of the first Fresnel zone has to be calculated. The maximum radius of the first Fresnel zone is calculated by determining the radius of the Fresnel zone cross section from Equation (3.4) at the midpoint between the transmitter and the receiver, as shown

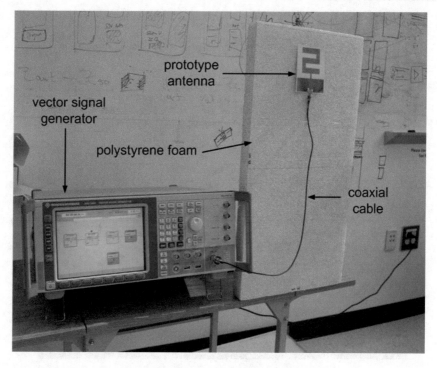

Figure 3.27: Prototype antenna (transmitter) and vector signal generator setup.

in Fig. 3.30.

$$r = \sqrt{\frac{\lambda}{\left(\frac{1}{d_t} + \frac{1}{d_r}\right)}} \qquad (3.4)$$

where:

r : the radius of the first Fresnel zone

λ : the wavelength of the transmitted signal in free space

d_t : the distance from the fresnel zone cross section to the transmitter

d_r : the distance from the fresnel zone cross section to the receiver

The radius r, when $d_t = d_r = 2.03$ m is found $r = 0.58$ m. It is observed that r is lower than the heights of the transmitter and the receiver and since no object exists between them, the requirement of clearance of the first Fresnel zone is satisfied. As a result the free space loss model can be applied as the propagation model in the wireless link of the two antennas.

Since the free space loss propagation model can be applied, the gain of each prototype antenna will be calculated using the Friis transmission equation [7]. The power received by an antenna is described by the Friis formula, which accounts for all antenna gains, path loss and losses in the

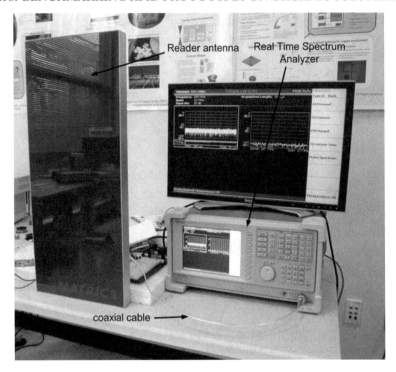

Figure 3.28: AN-400 RFID reader antenna (receiver) and real time spectrum analyzer setup.

system and is given for the current measurement by Equation (3.5)

$$\frac{P_r}{P_t} = G_t G_r \left(1 - |\rho|^2\right) \left|\hat{\rho}_t \cdot \hat{\rho}_r\right|^2 \left(\frac{\lambda}{4\pi d}\right)^2 \tag{3.5}$$

where:

P_t : the radiated power from the transmitter antenna

P_r : the received power from the receiver antenna

ρ : the complex reflection coefficient at the input of the transmit antenna

$\hat{\rho}_t$: polarization unit vector of the transmitter antenna

$\hat{\rho}_r$: polarization unit vector of the receiver antenna

  The Friis formula can be written in a decibel form, as shown in the equation below:

$$P_r = P_t + G_t + G_r - L_m - L_p - 20 \log_{10}\left(\frac{4\pi}{\lambda}\right) - 20 \log_{10}(d) \tag{3.6}$$

  It has to be noted that the polarization mismatch is given by $\left|\hat{\rho}_t \cdot \hat{\rho}_r\right|^2$ and equals the mismatch loss L_m in dB. From the above equation the gain for both the inkjet-printed and the copper tape antenna is calculated: $G_t = -0.33$ dBi.

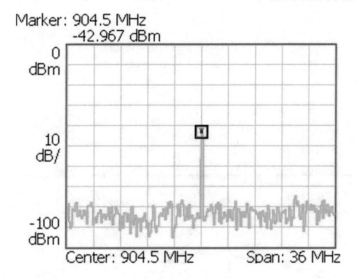

Figure 3.29: Received power at the AN-400 RFID reader antenna (receiver) terminals.

The obtained results, along with the results from the simulation are summarized in Table 3.1. In addition the design goals are also given for comparison: It has to be noted that the input impedance, directivity, gain, antenna efficiency and return loss are calculated at the required antenna operation frequency, at 904.5 MHz. Since no radiation pattern measurements were carried out, the directivity and the efficiency of the monopole antenna could not be experimentally measured.

It is observed that the return loss and gain measured results are in close agreement with the simulation and therefore the antenna design requirements defined in the first step of the design process are satisfied. It is seen from the measured −10 dB impedance bandwidth, that both the inkjet-printed and the copper tape fabricated antenna can operate efficiently across the USA (902 – 928 MHz) UHF RFID frequency band. The inkjet-printed antenna can operate also in the Europe (866 – 868 MHz) UHF RFID band, satisfying the design requirements.

As a result the feasibility of modeling, designing and efficiently printing antennas on paper substrate is verified. Paper proved to be an excellent candidate as an antenna dielectric substrate material, since its dielectric loss doesn't decrease significantly the antenna gain, as seen from the gain measurements results. The agreement of the inkjet-printed antenna measured results with the simulation, qualifies the inkjet printing process as an efficient method for printing RF structures, such as antennas on paper substrate.

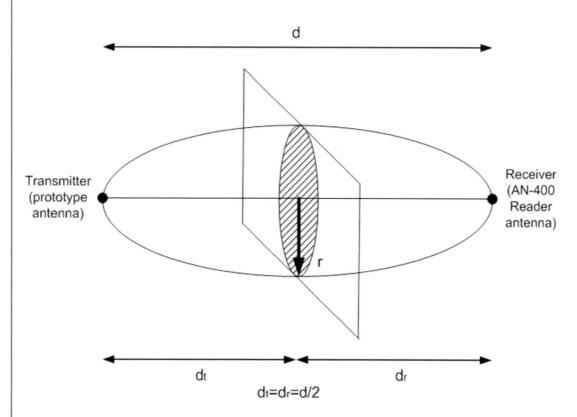

Figure 3.30: First Fresnel zone cross section at the middle point of the wireless link setup between the prototype antenna and the reader antenna (receiver).

Table 3.1: Comparison of simulated and experimental results of the proposed antenna with the design requirements

Antenna Parameter	Requirement	Simulation	Inkjet-Printed	Copper Fabricated
Center Frequency (MHz)	904.5	904	898	906
Operation Bandwidth (MHz)	$\geq 866 - 928$	$822 - 954$	$860 - 942$	$874 - 942$
Bandwidth (MHz)	≥ 62	132	82	68
Bandwidth (%)	≥ 6.9	14.6	9.1	7.5
Input Impedance (Ω)	$\approx 37.3 - j65.96$	$37.5 - j65.2$	$36.2 - j61.5$	$39.5 - j65.4$
Directivity (dBi)	≥ 0	0.23	-	-
Gain (dBi)	≥ -1	-0.42	-0.33	-0.33
Efficiency	≥ 80	88.7	-	-
Return Loss (dB)	≥ -15	-38.4	-23.7	-31.9

Bibliography

[1] K. Finkenzeller, RFID Handbook, 2nd ed., Wiley, 2004.

[2] P. V. Nikitin, S. Rao, S. F. Lam, V. Pillai, and H. Heinrich, "Power Reflection Coefficient Analysis for Complex Impedances *in RFID Tag Design*" IEEE Transactions on Microwave Theory and Techniques, Vol. 53, Sep. 2005. DOI: 10.1109/TMTT.2005.854191

[3] S. D. Kulkarni, R. M. Boisse, and S. N. Makarov, "*A Linearly-Polarized Compact UHF PIFA with Foam Support*", Department of Electrical Engineering, Worcester Polytechnic Institute.

[4] M. M. Tentzeris, L. Yang, A. Rida, A. Traille, R. Vyas, and T. Wu, "*RFID's on Paper using Inkjet-Printing Technology: Is it the first step for UHF Ubiquitous "Cognitive Intelligence" and "Global Tracking"?*" RFID Eurasia, 2007 1st Annual, pp. 1–4, 5-6 Sept. 2007. DOI: 10.1109/RFIDEURASIA.2007.4368098

[5] J. D. Griffin, G. D. Durgin, A. Haldi, and B. Kippelen, "RF Tag Antenna Performance on Various Materials Using Radio Link Budget," IEEE Antennas and Wireless Propagation Letters, Dec 2006. DOI: 10.1109/LAWP.2006.874072

[6] R. A. Burberry, "VHF and UHF Antennas," IEE Electromagnetic Waves Series 35. Peter Peregrinus Ltd. On behalf of the Institution of Electrical Engineers.

[7] C. Balanis, "Antenna Theory, Analysis and Design," 3rd ed., John Wiley & Sons, Inc., Publication.

CHAPTER 4

Conformal Magnetic Composite RFID Tags

As the technology for RFID systems continuously improves and extends to structures of non-orthogonal shapes and to conformal sensors of wireless body area networks (WBAN), there has been a need to design more "flexible" reader and tag systems. Namely miniaturization of the transponder and ability to tune the system performance to accommodate EM (electromagnetic) absorption and interference from surrounding media, while compensating for fabrication tolerances has been one of the major priorities [1]. Three-dimensional transponder antennas that utilize wound coil inductors do make use of magnetic cores, but they are quite bulky and impractical. On the other side, magnetic materials for two-dimensional embedded conformal planar antennas have not yet been successfully realized for standard use. This chapter introduces a novel flexible magnetic composite for printed circuits and antennas, that can reap the same miniaturization and tuning benefits with the heavier and non-flexible 3D used magnetic cores.

One of the most significant challenges for applying new magnetic materials is understanding the interrelationships of the properties of the new materials with design and performance of the specific topology (e.g., radiation pattern, scattering parameters). In previous studies, it can often be cited that the objectives of miniaturization and improved performance are tempered by the limited availability of materials that possess the required magnetic properties, while maintaining an acceptable mechanical and conformality performance [2]. Recently, formulation of nano-size ferrite particles has been reported [3] and formulation of magnetic composites comprised of ferrite filler and organic matrix has been demonstrated [4]. The implication of these new magnetic materials has yet not been investigated for specific EM systems above the low MHz range. Additionally, in the cases of complex microwave systems involving numerous interconnects, dielectric interfaces or radiating structures, the co-design of the materials along with the structure dimensions and the fabrication design rules may be necessary in order to achieve the optimal targeted performance. The aim of this work is to provide a basis for this co-design of materials, fabrication processes and electromagnetic structures, namely for the benchmarking case of a novel flexible magnetic composite, a BaCo ferrite-silicone composite, and a UHF RFID antenna, respectively.

Specifically, in this study a benchmark structure was first designed for 480 MHz in a full-wave simulator for an unfilled Silicone substrate; then the magnetic nano-particles were added and the same antenna was redesigned for 480 MHz by reducing its size, thus proving the miniaturization concept. The next step was the actual fabrication of the material and the measurement of the electrical and magnetic characteristics, including loss. Finally, the miniaturized antenna was fabricated on the magnetic composite and its performance was measured, compared and validated along with the simulated predictions. The presented magnetic substrate is the first flexible magnetic composite tested and proven for the 480 MHz bandwidth with acceptable magnetic losses, which makes it usable for lightweight conformal/wearable applications like pharmaceutical industry and wireless health monitoring in hospital, ambulance and home-based patient care.

The first step for this work was to develop a magnetic composite that provides the advantage of low temperature processing for compatibility with organic substrate processing, flexibility, and high adhesion. With regard to these three properties, the magnetic composite would have to be compatible with common substrates used for RFID, such as polyethylene terephthalate (PET) and polyimide. Additionally, the composite dielectric loss can affect circuit performance, and low dielectric loss would be targeted. For these objectives, properties of candidate matrices would be the same, that is, low temperature processibility, high flexibility, high adhesion, and low dielectric loss. Dielectric constant can also affect the circuit performance and should also be monitored. The matrix materials considered candidates for this proposed work included silicone, UV curable acrylic-based adhesives, and benzocyclobutene (BCB). Silicone provides reasonable viscosities required for good filler mixing during processing, that is, not too low to promote settling and not too high for uniform mixing. Additionally, silicone provides the properties of flexibility and, for some formulations, good adhesion.

After careful analysis, the matrix material choice was made for Dow Corning Sylgard 184 silicone. The electrical parameters of the unfilled silicone, used in the initial antenna design, are $\varepsilon_r = 2.65$ and $\tan \delta_e = 0.001$. The choice for the magnetic composite was Trans-Tech BaCo ferrite powder, product name Co_2Z. A 40 vol % ferrite paste was produced with a mixer at 240 rpm and 110°C for 30 minutes. The paste was transferred into a flat mold and vacuum cured with a hold confirmed to occur at >125°C for 50 minutes to produce a 1.3 mm thick substrate.

The material was measured using an HP4291A impedance analyzer to obtain complex permittivity (ε) and permeability (μ) (real and imaginary parts) with material fixtures 16453A for ε and 16454A for μ over the frequency range of 1 MHz to 1.8 GHz. There were 5 measurements taken for each ε_r, μ_r, $\tan \delta_e$ and $\tan \delta_m$. The summary statistics, including the mean and 95% C.I. (confidence intervals) for ε_r, μ_r, $\tan \delta_e$ and $\tan \delta_m$ of the ferrite composite at 480 MHz are given in Table 4.1. Based on these results, the values used in the model were $\varepsilon_r = 7.14$, $\mu_r = 2.46$, $\tan \delta_e = 0.0017$ and $\tan \delta_m = 0.039$.

The free space wavelength at 480 MHz is 692 mm. An RFID tag that has the miniaturized features is becoming more of a necessity, for example, in the implementation of an RFID-enable wristband for wireless health monitoring in hospital. To achieve this design goal, a folded bow-tie meander line dipole antenna was designed and fabricated on the characterized magnetic composite material substrate. The RFID prototype structure is shown in Fig. 4.1 along with dimensions, with the IC placed in the center of the shorting stub arm. The nature of the bow-tie shape of the half-wavelength dipole antenna body allows for a broader band operation. The meander line helps realizing further miniaturization of the antenna structure. The shorting stub arm is responsible for the matching of the impedance of the antenna terminals to that of the IC through the fine tuning of the length. Fig. 4.2 shows the resistance and reactance versus frequency when the shorting stub arm is tuned at 12 mm, 9 mm, and 6.8 mm, respectively.

	mean	Lower CI	Upper CI
Table 4.1: Mean and 95% Confidence Intervals for Measurements of Ferrite Composite at 480 MHz			
ε_r	7.142	7.083	7.201
μ_r	2.463	2.457	2.468
$\tan \delta_e$	0.0017	0.0005	0.0028
$\tan \delta_m$	0.0391	0.0358	0.0424

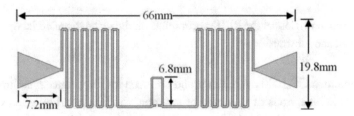

Figure 4.1: Configuration of the RFID tag module on magnetic composite substrate.

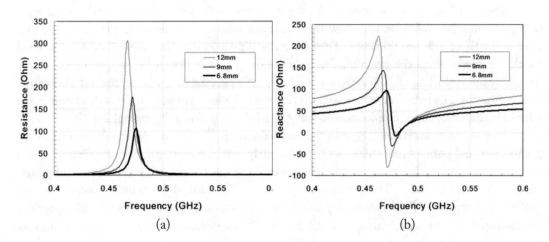

Figure 4.2: Simulated input resistance and reactance of the RFID tag with the shorting stub length of 12 mm, 9 mm, and 6.8 mm, respectively. (a) Resistance (b) Reactance.

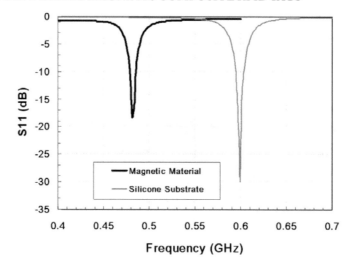

Figure 4.3: Measured return loss of the RFID tag antenna on the magnetic material with the comparison of the one on the silicone substrate.

In measurement, a GS 1000 μm pitch probe was used for impedance measurements. In order to minimize backside reflections of this type of antenna, the fabricated antenna was placed on a custom-made probe station using high density polystyrene foam with low relative permittivity of value 1.06 resembling that of the free space. The calibration method used was short-open-load-thru (SOLT).

The initial structure was designed for the lower end of the UHF spectrum and was modeled using Zeland IE3D full wave EM software. The initial substrate was pure silicone ($\varepsilon_r = 2.65$ and $\tan \delta = 0.001$) of 1.3 mm thickness. Then the same dimensions of the antenna were maintained for the magnetic composite material. The Return Loss plot is shown in Fig. 4.3, demonstrating a frequency down shifting of 20% with increased magnetic permeability, which proves the miniaturization concept. The radiation pattern of the RFID tag module was plotted in Fig. 4.4. The radiation pattern is almost uniform (omnidirectional) at 480 MHz with directivity around 2.0 dBi.

One of the most critical factors in the magnetic composite fabrication was the control of the permittivity and permeability values, so a careful analysis of the impact of both the dielectric and magnetic performance based on the fabrication variability was necessary. These material properties are not mutually exclusive. The permittivity [5] and permeability [6] are both governed by the molecular arrangement (lattice structure) and elemental composition of the material, which prevents the tuning of these properties independently. So the following analysis does not attempt to optimize the material parameters, but rather to quantify the effect of the parameters on the system-level performance of the antenna.

First, the impact of the loss tangents was investigated. The methodology used involves electromagnetic simulations and statistical tools and is presented as a flowchart in Fig. 4.5. First, Design

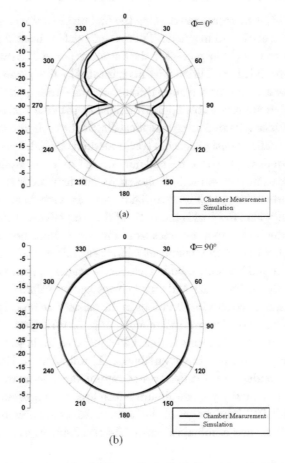

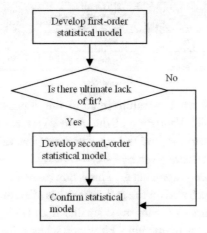

Figure 4.4: Simulated vs. measured 2D radiation plots for (a) $\Phi = 0°$ and (b) $\Phi = 90°$.

Figure 4.5: Procedure for statistical model development.

of Experiments (DOE) [7] is performed to develop the first order (linear) statistical model, including both loss tangents, dielectric and magnetic. Then, the model is checked for ultimate lack of fit, more specifically, if curvature might be present in the output response. If curvature in the response is detected, the analysis is extended to additional points indicated by the Response Surface Methodology (RSM) [7] which can account for curvature through second-order model development. Usually, these second-order models are reasonable approximations of the true functional relationship over relatively small regions. Once validated using statistical diagnostic tools, the models approximate the actual system within the defined design space. Hybrid methods including statistical tools and EM simulations have been extensively used for RF and microwave systems analysis and optimization [8].

The statistical experimentation method chosen for the first-order statistical model is a full factorial DOE with center points [7]. The factorial designs are used in statistical experiments involving several (k) factors where the goal is the study of the joint effects of the factors on a response and the elimination of the least important ones from further optimization iterations. The $2k$ factorial design is the simplest one, with k factors at 2 levels each. It provides the smallest number of runs for studying k factors and is widely used in factor screening experiments [7]. Center points are defined at the center of the design space and enable investigating validity of the model, including curvature in the response, and account for variations in the fabrication process of the structure. Since the statistical models are based on deterministic simulations, the variations of the center points were statistically simulated assuming a 3 process with a $\pm 2\%$ tolerance for both $\tan \delta_e$ and $\tan \delta_m$.

In this case, since we have two input variables, a 22 full factorial DOE was performed for the first-order statistical model, with the following 4 output variables as the antenna performance figures of merit: resonant frequency f_{res}, minimum return loss RL, maximum gain at 480 MHz G, and the 10 dB bandwidth BW. The ranges of the input variables are presented in Table 4.2, while ε_r and μ_r have been kept at their nominal values of 7.14 and 2.46, respectively.

Table 4.2: Ranges for the Input Variables			
Variable	**Low value "−"**	**High value "+"**	**Center point**
$\tan \delta_e$	0.00136	0.00204	0.0017
$\tan \delta_m$	0.0312	0.0468	0.039

The first-order models showed curvature in all of the responses, and RSM was needed for the second-order statistical model. Validation of the models was investigated, with all but the BW validated for the normality assumption, and the equal variance was validated for RL and G, but not for f_{res} and BW. The four models are given by (4.1)–(4.4). An interesting result is the fact that the resonant frequency is not dependent upon $\tan \delta_e$. This is due to the fact that the interval of analysis of $\tan \delta_e$ shown in Table 4.2, chosen based on the real material, is of an order of magnitude smaller than $\tan \delta_m$, because $\tan \delta_e$ is of an order of magnitude smaller than $\tan \delta_m$ and the intervals are chosen to be 20% up and down the center point value. However, when reflected in loss and bandwidth in

Equations (4.2)–(4.4), even the much smaller $\tan \delta_m$ becomes significant.

$$f_{res}(MHz) = 480.47 - 0.024 \left(\frac{\tan \delta_m - 0.039}{0.0078} \right) - 0.013 \left(\frac{\tan \delta_m - 0.039}{0.0078} \right)^2 \tag{4.1}$$

$$RL(dB) = -20.34 + 0.18 \left(\frac{\tan \delta_e - 0.0017}{0.00034} \right) + 2.48 \left(\frac{\tan \delta_m - 0.039}{0.0078} \right)$$

$$- 0.062 \left(\frac{\tan \delta_m - 0.039}{0.0078} \right)^2 - 0.43 \left(\frac{\tan \delta_e - 0.0017}{0.00034} \right) \left(\frac{\tan \delta_m - 0.039}{0.0078} \right) \tag{4.2}$$

$$G(dBi) = -4.57 - 0.019 \left(\frac{\tan \delta_e - 0.0017}{0.00034} \right) - 0.26 \left(\frac{\tan \delta_m - 0.039}{0.0078} \right)$$

$$- 0.0044 \left(\frac{\tan \delta_m - 0.039}{0.0078} \right)^2 + 0.0005 \left(\frac{\tan \delta_e - 0.0017}{0.00034} \right) \left(\frac{\tan \delta_m - 0.039}{0.0078} \right) \tag{4.3}$$

$$BW(MHz) = 7.69 + 0.0000088 \left(\frac{\tan \delta_e - 0.0017}{0.00034} \right) + 0.038 \left(\frac{\tan \delta_m - 0.039}{0.0078} \right)$$

$$- 0.031 \left(\frac{\tan \delta_m - 0.039}{0.0078} \right)^2 - 0.005 \left(\frac{\tan \delta_e - 0.0017}{0.00034} \right) \left(\frac{\tan \delta_m - 0.039}{0.0078} \right) \tag{4.4}$$

The models allow for the "a priori" prediction of the antenna performance with respect to either figure of merit or all simultaneously allocating any weight factors to each one of them. The goals chosen in this case were a specific f_{res} of 480 MHz (center point value), maximum gain G, minimum return loss RL, and maximum bandwidth BW, all with equal weight. The surfaces for the four figures of merit as a function of the input parameters are presented in Fig. 4.6, indicating the curvature in the models. The values that satisfied the four goals within the ranges presented in Table 4.2 were $\tan \delta_e = 0.00136$ and $\tan \delta_m = 0.032427$, leading to the predicted values of the four figures of merit of $f_{res} = 480.48$ MHz, $RL = -22.97$ dB, $G = -4.32$ dBi and $BW = 7.63$ MHz. So, ideally, these values of the loss tangents would provide optimal performance of the antenna for the above mentioned goals. The models indicate that the resonant frequency decreases with the losses, as the gain and the return loss obviously degrade. For the bandwidth, although the model is significant and shows an increase of the bandwidth with dielectric loss, the absolute numbers in the RSM vary only between 7.61 and 7.7 MHz, which is not a large difference for practical applications.

The consideration of the relative permeability in the antenna design requires a more detailed analysis of its impact, together with the relative permittivity, on the antenna performance. The next statistical experiment analyzes the impact of these two parameters on the same major antenna outputs: resonant frequency f_{res}, minimum return loss RL, maximum gain at the resonant frequency G, and the 10 dB bandwidth BW.

The methodology used is the same as the one used for the loss tangent analysis and shown in Fig. 4.5.

$$f_{res}(MHz) = 480.61 - 15.19 \left(\frac{\varepsilon_r - 7.14}{1.428} \right) - 9.5 \left(\frac{\mu_r - 2.64}{0.492} \right)$$

$$+ 1.01 \left(\frac{\varepsilon_r - 7.14}{1.428} \right)^2 + 1.46 \left(\frac{\mu_r - 2.64}{0.492} \right)^2 \tag{4.5}$$

$$RL(dB) = -20.34 + 1.19 \left(\frac{\varepsilon_r - 7.14}{1.428} \right) - 2.78 \left(\frac{\mu_r - 2.64}{0.492} \right)$$

$$+ 0.39 \left(\frac{\varepsilon_r - 7.14}{1.428} \right) \left(\frac{\mu_r - 2.64}{0.492} \right) - 0.14 \left(\frac{\varepsilon_r - 7.14}{1.428} \right)^2 - 0.11 \left(\frac{\mu_r - 2.64}{0.492} \right)^2 \tag{4.6}$$

$$G(dBi) = -4.56 - 0.19 \left(\frac{\varepsilon_r - 7.14}{1.428} \right) + 0.032 \left(\frac{\mu_r - 2.64}{0.492} \right)$$

$$+ 0.0087 \left(\frac{\varepsilon_r - 7.14}{1.428} \right)^2 + 0.0024 \left(\frac{\mu_r - 2.64}{0.492} \right)^2 \tag{4.7}$$

$$BW(MHz) = 7.68 - 0.35 \left(\frac{\varepsilon_r - 7.14}{1.428} \right) - 0.09 \left(\frac{\mu_r - 2.64}{0.492} \right)$$

$$+ 0.037 \left(\frac{\varepsilon_r - 7.14}{1.428} \right) \left(\frac{\mu_r - 2.64}{0.492} \right) + 0.022 \left(\frac{\varepsilon_r - 7.14}{1.428} \right)^2 - 0.029 \left(\frac{\mu_r - 2.64}{0.492} \right)^2 \tag{4.8}$$

In this case, since we have two input variables, the same 22 full factorial DOE was performed for the first-order statistical model, with the ranges of the input variables presented in Table 4.3, while $\tan \delta_e$ and $\tan \delta_m$ have been kept at their nominal values of 0.0017 and 0.039, respectively.

Table 4.3: Ranges for the Input Variables			
Variable	Low value "−"	High value "+"	Center point
ε_r	5.712	8.568	7.14
μ_r	1.968	2.952	2.46

The first order models showed curvature in all of the responses, and RSM was needed for the second-order statistical model. The validation of the models was investigated. For the normality of residuals assumption, all models but G have normally distributed residuals. For the validation of the equal variance of residuals assumption, all the models had equal variance of residuals.

The antenna performance was predicted again for the same goals: goals: f_{res} of 480 MHz (center point value), maximum gain G, minimum return loss RL, and maximum bandwidth BW, all with equal weight. The surfaces for the four figures of merit as a function of the input parameters are presented in Fig. 4.7, indicating slight curvature in the models. The values that satisfied the

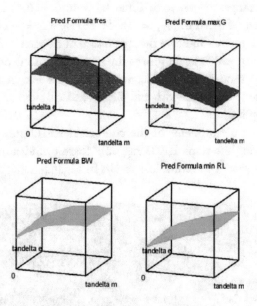

Figure 4.6: Surfaces of possible solutions for outputs.

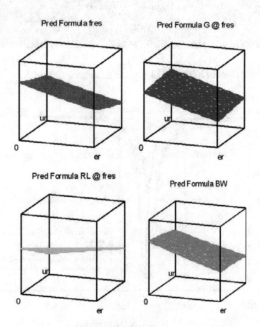

Figure 4.7: Surfaces of possible solutions for outputs.

four conditions within the ranges presented in Table 4.3 were $\varepsilon_r = 6.41$ and $\mu_r = 2.95$, leading to the values of the four figures of merit of $f_{res} = 480.56$ MHz, $RL = -24.08$ dB, $G = -4.42$ dBi and $BW = 7.73$ MHz. The models indicate that the resonant frequency decreases with the relative permittivity and permeability, which again proves the miniaturization concept. For the bandwidth and the gain, although the models are significant, the absolute numbers in the RSM vary only between $7.23 - 8.22$ MHz for the bandwidth and $4.27 - 4.81$ dBi for the gain, which are not large differences for practical applications.

In order to verify the performance of the conformal RFID antenna, measurements were performed as well by sticking the same RFID tag on a foam cylinder, as shown in Fig. 4.8. The radius of the cylinder is 54 mm. The return loss results in Fig. 4.9 show that the return loss of the

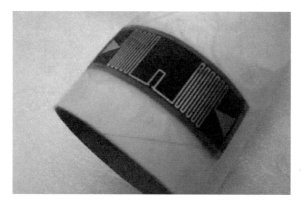

Figure 4.8: Photograph of the conformal RFID tag on a foam cylinder.

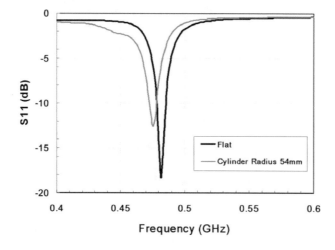

Figure 4.9: Measured return loss of the flat RFID tag and the conformal RFID tag.

inkjet-printed antenna is slightly shifted down by 6 MHz with a center frequency at 474 MHz. Overall a good performance is still remained with the interested band covered. The flexible property of the substrate enables the RFID tag module's application in diverse areas. Fig. 4.10 demonstrates the conformal RFID tag prototype in the applications of wireless health monitoring and pharmaceutical drug bottle tracking.

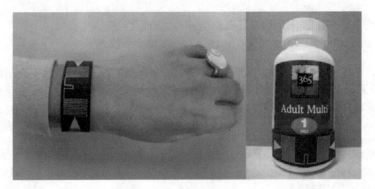

Figure 4.10: Embodiments of the conformal RFID tag prototype in the applications of wireless health monitoring and pharmaceutical drug bottle tracking.

This work is the first demonstration of a flexible magnetic composite proven for the 480 MHz bandwidth with acceptable magnetic losses that makes it usable for small size, lightweight conformal applications like wireless health monitoring in pharmaceuticals, hospital, ambulance and home-based patient care. A combination of electromagnetic tools and measurements has been used to investigate the impact of magnetic composite materials to the miniaturization of RFID antennas considering geometric and material parameters, as well as conforming radius. This approach has been applied to the design of a benchmarking conformal RFID tag module and has enabled the assessment of implication that the choice of materials have on this design, specifically the antenna miniaturization by using the magnetic composite vs. pure silicone. A real composite material has been fabricated and the performance of the miniaturized antenna predicted using the models. Next, the important issues of the dielectric and magnetic losses has been addressed by performing a thorough statistical analysis to investigate the impact of the losses on the antenna performance. Furthermore, since the permeability was first introduced in this paper for a conformal antenna, the impact of the relative permeability in conjunction with relative permittivity were addressed together in another statistical analysis. The losses impact the resonant frequency, return loss, and antenna gain, whereas the dielectric constant and magnetic property mostly decrease the resonant frequency, thus proving the miniaturization concept.

Bibliography

[1] "Magnetic Materials for RFID," TechnoForum 2005, TDK, http://www.tdk.co.jp/tf2005/pdf_e/2f0215e.pdf.

[2] N. Das and A. K. Ray, "Magneto Optical Technique for Beam Steering by Ferrite Based Patch Arrays," IEEE Transactions on Antennas and Propagation, vol. 49, no. 8, pp. 1239–1241, August 2001. DOI: 10.1109/8.943321

[3] S. Morrison, C. Cahill, E. Carpenter, S. Calvin, R. Swaminathan, M. McHenry, and V. Harris, "Magnetic and Structural Properties of Nickel Zinc Ferrite Nanoparticles Synthesized at Room Temperature," Journal of Applied Physics, vol. 95, pp. 6392–6395, June (2004). DOI: 10.1063/1.1715132

[4] H. Dong, F. Liu, Q. Song, Z.J. Zhang, and C. P. Wong, "Magnetic Nanocomposite for High Q Embedded Inductor," IEEE International Symposium and Exhibition on Advance Packaging Materials: Process, Properties, and Interfaces, Atlanta, Georgia, pp. 171–174, 2004. DOI: 10.1109/ISAPM.2004.1288008

[5] S. O. Kasap, Principles of Electronic Materials and Devices, 2nd ed., The McGraw-Hill Companies, New York, p. 516, 2002.

[6] L. L. Hench and J. K. West, Principles of Electronic Ceramics, John Wiley & Sons, New York, p. 296, 1990.

[7] J. Neter et al., "Applied Linear Statistical Models," 4th ed., The McGraw-Hill Companies, Chicago, 1996.

[8] D. Staiculescu, C. You, L. Martin, W. Hwang, and M. M. Tentzeris, "Hybrid Electrical/Mechanical Optimization Technique Using Time-Domain Modeling, Finite Element Method and Statistical Tools for Composite Smart Structures," Proc. of the 2006 IEEE IMS Symposium, pp. 288–291, June 2006. DOI: 10.1109/MWSYM.2006.249490

CHAPTER 5

Inkjet-Printed RFID-Enabled Sensors

As the demand for low cost, flexible and power-efficient broadband wireless electronics increases, the materials and integration techniques become more and more critical and face more challenges, especially with the ever growing interest for "cognitive intelligence" and wireless applications, married with RFID technologies. This demand is further enhanced by the need for inexpensive, reliable, and durable wireless RFID-enabled sensor nodes that is driven by several applications, such as logistics, Aero-ID, anti-counterfeiting, supply-chain monitoring, space, healthcare, pharmaceutical, and is regarded as one of the most disruptive technologies to realize truly ubiquitous ad-hoc networks. Hence, the two major challenges for such applications are the choice of the material and the advanced integration capabilities. The choice of paper or LCP as the substrate material presents multiple advantages and has established the organic substrate as the most promising materials for UHF RFID applications: their high biodegradability with respect to other ceramic substrates such as FR-4, requiring only months to turn into organic matter in land-fills. Previous work has demonstrated the successful development of fully inkjet-printed RFID modules on paper [1]. The next challenge is to integrate the sensor on the paper substrate as well.

From the power consumption prospect, RFID-enabled sensors can be divided into two categories: active ones and passive ones. Active RFID-enabled sensors have longer communication rage and stronger sensing capability assisted by the power from a battery. While passive RFID-enabled sensors enjoy the advantages of low-profile structure and maintenance free fashion.

In this chapter, the process to integrate electronics, IC's and sensors on paper using inkjet printed technology is presented. The design process from the system down to the fabrication will be outlined. The constitutive parameters of the paper reported in [1, 2] and [3] will be used to design the microwave structures in the wireless modules. System-level measurements highlighting the feasibility of paper as a suitable low-cost integration platform/package for wireless sensing applications will be demonstrated through wireless link measurements.

5.1 ACTIVE RFID-ENABLED SENSOR

To investigate the feasibility of paper as an RF frequency substrate and packaging material, two microcontroller-enabled wireless sensor modules were realized on a paper substrate. The system level design for the wireless transmitters can be seen in Fig. 5.1. At the heart of the system was an 8-bit microcontroller unit (MCU) integrated with a UHF ASK transmitter in a single integrated circuit chip (IC). Having an integrated MCU with a transmitter offered the versatility of customizing the system for use with any kind of analog or even digital sensor, whose data could be sent out wirelessly with a maximum user-defined data rate. The data transmission was to be carried out at the unlicensed Industrial Scientific and Medical frequency band around 900 MHz in a way similar to RFID tags that use EPC Global's GEN-2 protocol [4]. The first wireless sensor module prototype using a dipole antenna was printed on a 2-D (single layer) paper module, and the second prototype using a monopole antenna was printed on a 3-D (multilayer) paper module. The full design process will be outlined in this section.

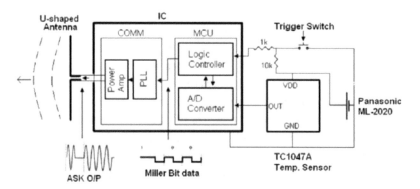

Figure 5.1: System Level Diagram of the Dipole and Monopole based Wireless Sensor Modules.

An integrated 8-bit Microcontroller unit (MCU) was used as the primary controlling mechanism of the wireless sensor module. The MCU used a high-performance Reduced Instruction Set Computing (RISC) architecture, with its own internal 4 MHz oscillator that was calibrated to within 1%. The MCU was programmed using Assembly code, which offers the best control over the timing of the device instructions and hence the I/O ports of the MCU. The MCU was programmed to operate in three different modes of operation, which could be selected by the user with an external trigger switch (push button) shown in the system level diagram in Fig. 5.1. In mode 1 (UNMOD), the MCU was programmed to turn on the COMM module, and have it send out an un-modulated signal at the design frequency around 900 MHz. In mode 2 (SENSE), the MCU was programmed to read temperature from an external temperature sensor and have the COMM module send out the sensor data using an encoded ASK modulated signal. In mode 3, the MCU was programmed to disable the COMM module and go into sleep mode (SLEEP), an extremely low power state (1.8 μWatts), in order to conserve power [5, 6]. In the UNMOD mode, during the full unmodulated signal transmission, the power consumption was measured to be 36 μWatts, which would enable the wireless sensor module system to operate for a maximum period of 3.75 hours with a 3V battery capable of supplying 45 mA-hours [7].

The MCU contained an integrated 10-bit Analog-to-Digital (A/D) converter that was setup to sample a Microchip TC1047A analog temperature sensor in the SENSE mode. All 10 bits of the A/D converter were utilized for maximum sensing resolution. The temperature sensor was sampled

4 times during one read operation to average out variations in the sampled sensor output in-between its sensor readings. The sampled analog data were then converted to their respective 10-bit digital form and stored in the MCU's program memory, where they were averaged. In assembly code the averaging was performed by summing all four of the stored digital sensor data and performing a bit shift operation on the sum twice to perform a divide by four operations.

The averaged sensor data in its digital form was then bit encoded using a complete 2 sub-carrier cycle Miller bit encoding. Miller bit encoding is one of the most common symbol types used in RFID's GEN-2 protocol [8]. In it, 2 or 3 bit transitions occur in between a '1' or '0' symbol, respectively, and no two consecutive 0 or 1 bits can be identical. This makes it possible to communicate the bits asynchronously making it easier to decode them on the receiver side, while conserving bandwidth. As a final step, the bit-encoded sensor data was then transmitted wirelessly by using the MCU to control the integrated Communication module (COMM).

The transmit frequency was generated with the help of an external crystal oscillator fed into the communication module (COMM) of the IC in Fig. 5.1, which comprised of a Phase Lock Loop (PLL) module with a Power Amplifier (PA) at the output of the wireless transmitter. The COMM module, when enabled by the MCU, takes in the input or reference frequency generated by the RF crystal oscillator and generates an output signal locked in at 32 times the reference frequency. The MCU was programmed to produce a clock delay of 600 micro-seconds after enabling the integrated COMM module, which is roughly equal to the startup time of the crystal oscillator and the acquisition time of the PLL. The acquisition time is the amount of time taken by the PLL to lock at which point the output of the phase/freq. detector and the Voltage controlled oscillator block in Fig. 5.2 are 0A and 0V, respectively [9]. The PLL lock occurs only when its output frequency is equal to 32 times the reference frequency of the RF crystal oscillator.

Half-wavelength dipole antennas are selected for the RF-front end, since they are among the most commonly used antennas for RFID applications. They can be folded and tapered for more compact size and wider bandwidth [10, 11]. In addition, half-wavelength dipole antennas have a radiation resistance of 75 ohms [12], which is close to the real part of the optimum load impedance looking out of the PA at 904.4 MHz making for an easier impedance match. For the first wireless sensor module prototype, a tapering U-shaped half wavelength dipole structure described in [13] was chosen for the antenna. The circuit layout was to be placed in the space within the U-shape of the antenna to make the size of the complete module more compact. The antenna was to be printed on a single layer of a paper substrate along with the circuit layout using inkjet printing technique. The biggest challenge on using a dipole antenna this way was integrating and sufficiently isolating it from the circuit of the sensor module within its U-shape, as shown in Fig. 5.3. The close vicinity of the antenna with respect to the circuit would alter its mutual impedance, which could cause an impedance mismatch with the PA thereby reducing the amount of power radiated along with the transmitted range [12]. To ensure an impedance match, the antenna was optimized so that the impedance looking out of the PA would be close to its optimum value at 904.4 MHz (60.1−j73.51 ohms) with the sensor module circuitry connected within the space between the antenna arms.

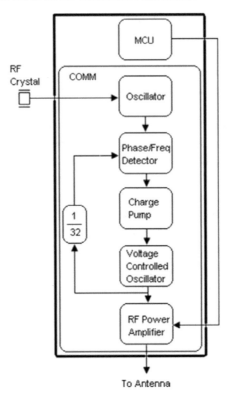

Figure 5.2: System Level Diagram of Wireless transmitter.

In addition, RF chokes (L1, L2) were used at several points in the circuit to minimize RF signals, meant to travel between the antenna and PA, from creeping into the positive power supply trace of the battery that was used to bias the PA (antenna terminals) and power up the MCU. RF chokes could not however be placed between the negative terminal of the antenna (also the PA ground) and the negative power supply of the battery because doing so would reduce the gain of the single ended PA and alter the optimum load impedance determined [14]. An RF choke (L3) was placed between the RF crystal and the negative power supply of the battery to isolate it from RF signals from the antenna. The entire structure of the module including the antenna and the circuit layout was built using Ansoft's HFSS 3-D EM solver, which was also used to optimize the antenna. Lumped RLC (Resistor /Inductor/ Capacitor) boundaries with values equal to the measured values of inductors L1, L2, and L3 at the 904.4 MHz were used to simulate the RF chokes in the circuit, as shown in Fig. 5.3. A lumped port was used as the RF power source in place of the PA to excite the antenna. The final dipole antenna design along with the circuit in between had a dimension of 9.5 cm by 5 cm. The simulated return loss or S11 for the antenna with respect to the PA's optimum load impedance (60.1−j73.51 ohms) is shown in Fig. 5.4, and shows a good impedance match at the

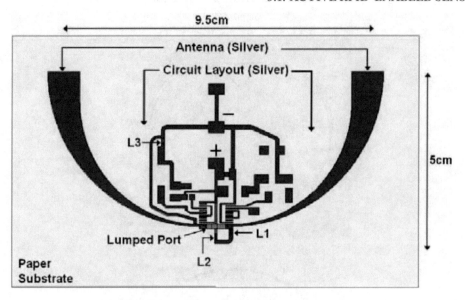

Figure 5.3: Dipole based Module topology.

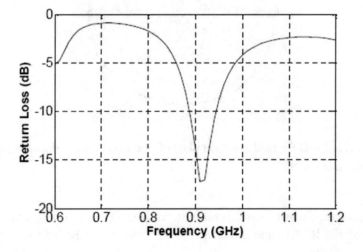

Figure 5.4: Simulated Return loss of the Dipole antenna connected to the circuit.

design frequency of 904.4 MHz with a −10 dB bandwidth of 60 MHz. The simulated and measured radiation patterns are shown in Fig. 5.5 and show reasonably good agreement. A maximum simulated directivity of 1.54 dB was achieved.

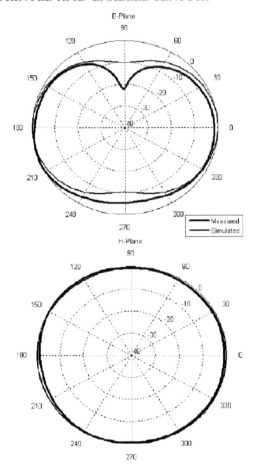

Figure 5.5: Normalized 2-D far field radiation plots of simulation and chamber measurement of the dipole based printed sensor module.

The assembled dipole based wireless sensor module can be seen in Fig. 5.6. The transmitted signal measured by the RTSA from a distance of 4.26 meters can be seen in Fig. 5.7, and was observed to be −48.07 dBm.

In order to verify the correct operation of the dipole-based wireless sensor module, it was triggered to operate in the SENSING mode. The ASK modulated sensor information sent out by the module at different temperatures that was measured by the RTSA is shown in Fig. 5.8. The transmitted sensor data shows good agreement with the measurements carried out with the digital infrared thermometer. The digital infrared thermometer has an accuracy of ±2.5 Degree Celsius [15].

Many of the drawbacks with the dipole based module can be eliminated by using a monopole based structure. The monopole uses its ground planes as a radiating surface, which can also be used

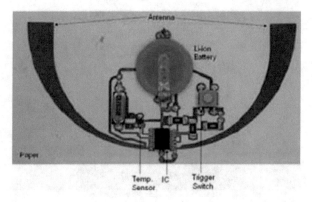

Figure 5.6: Dipole based wireless sensor module on paper substrate using inkjet printing technology.

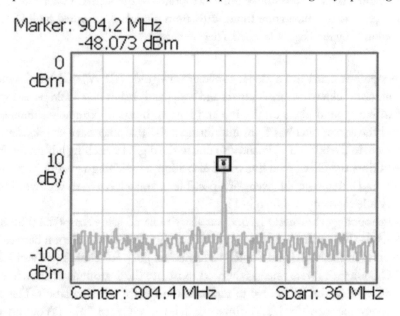

Figure 5.7: RTSA measured ASK modulated signal strength for the dipole based module from a distance of 4.26 meters.

to shield any circuitry behind it. The monopole antenna also does not require a differentially fed input signal like the dipole, which was ideal for the PA since its output was single-ended.

The circuit for the monopole was laid out on 2 layers, which helped minimize the size of the circuit topology by avoiding the long power supply traces that had to be used on the single layer dipole based module. The top layer contained the printed antenna and most of the circuit components for the module. The bottom layer contained a Li-ion cell and the power supply traces, which were routed to the top layer through drilled vias.

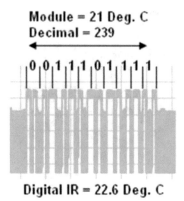

Figure 5.8: ASK modulated temperature sensor data captured by the RTSA at room temperature (Power vs. Time). Module: Sensed Temperature transmitted from module & captured by RTSA; Digital IR: Temperature measured by the Digital Infrared Thermometer.

The monopole antenna had a planar coplanar waveguide (CPW) wideband structure with a rectangular radiator to achieve a more compact and wideband design that could be easily printed [16, 17]. A CPW with a ground plane on the top and bottom layers is extremely suitable at shielding the antenna and the sensor data bus from interfering noise that may have coupled into the shared power supply traces in the bottom layer and also due to the digital switching within the MCU on the top layer. In addition, the CPW feed line could also allow a matching network to be implemented between the PA and antenna in the event of a possible mismatch between the two. The monopole based sensor module topology is shown in Fig. 5.9.

The entire topology, shown in Fig. 5.9, was also simulated using Ansoft's HFSS 3D EM tool. Multipoint grounds (RF and LF) were used for this design for better isolation between the digital switching occurring in the MCU and the RF transmission [18]. RF chokes (L1 and L2) simulated as lumped RLC boundaries were once again used to isolate the 2 grounds as shown in Fig. 5.9. A lumped port was used as the RF source to replicate the PA for the simulation. The antenna was matched to an impedance of $60.1-j73.51$ ohms which is the reference at the PA output at the design frequency of 904.4 MHz. The simulated return loss for the entire structure showed good wideband resonance of about 220 MHz around the design frequency of 904.4 MHz as shown in Fig. 5.10. The maximum simulated directivity obtained was 2.6 dB. The measured and simulated radiation patterns are shown in Fig. 5.11.

The assembled monopole-based wireless sensor module can be seen in Fig. 5.12. Wireless link measurements were carried out with the monopole based wireless sensor module by placing them at different temperatures. The measurement setup was similar to the one used for the dipole based modules. The transmitter and the receiver antennas were placed at 1.83 meters apart. The transmitted signal measured by the RTSA can be seen in Fig. 5.13, and was observed to be -26 dBm at a frequency of 904.4 MHz.

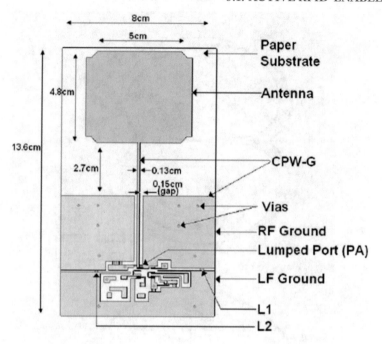

Figure 5.9: Monopole based module topology.

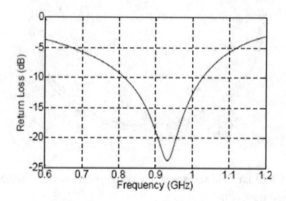

Figure 5.10: Simulated Return loss of the Monopole antenna connected to the circuit.

In order to verify the correct operation of the monopole based wireless sensor module, it was placed at different temperatures while triggered to operate in the SENSING mode. The ASK modulated sensor information sent out by the module at different temperatures that was measured by the RTSA is shown in Fig. 5.14. The transmitted sensor data shows good agreement with the measurements carried out with the digital infrared thermometer.

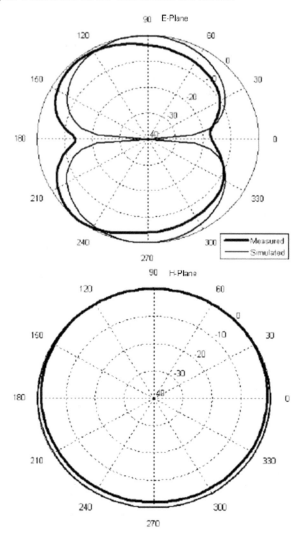

Figure 5.11: Normalized 2-D far field radiation plots of simulation and chamber measurement of the monopole based printed sensor module.

5.2 PASSIVE RFID-ENABLED SENSOR

The active RFID-enabled sensor tags use batteries to power their communication circuitry, and benefit from relatively long wireless range. However, the need of external battery limits their applications to where battery replacements are only possible and affordable. Battery technology is mature, extensively commercialized, and completely self-contained. However, given current energy density and shelf-life trends, even for relatively large batteries and conservative communication schedules,

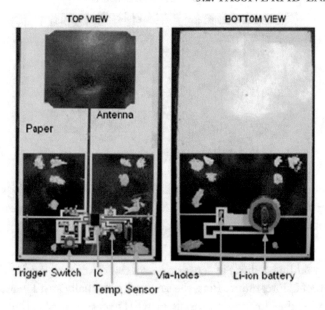

Figure 5.12: Monopole based wireless sensor module.

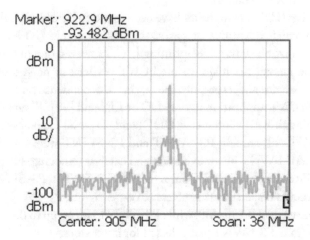

Figure 5.13: RTSA measured ASK modulated signal strength for the monopole based module.

the mean time to replacement is only a year or two. For some applications, such as harsh environment monitoring, in which battery changing is not easy, the problem is aggravated significantly. Concerns over relatively short battery life have restricted wireless device applications.

Researchers have been looking for passive RFID-enabled sensor solutions. Some passive RFID prototypes for sensing applications have been proposed [19, 20]. However, the sensing capabilities

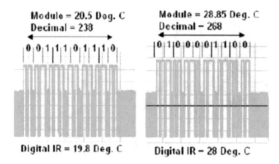

Figure 5.14: ASK modulated temperature sensor data captured by the RTSA. Module: Sensed Temperature transmitted from module & captured by RTSA; Digital IR: Temperature measured by the Digital Infrared Thermometer.

are usually realized by adding a discrete sensor or a special coating to the RFID tag, resulting in the difficulty in low-profile integration. Plus, the sensitivity is usually low. Therefore, there has been a growing interest in looking for new materials in RFID sensing applications: an ultra sensitive composite which can be printed directly on the same paper together with the antenna, for a low cost, flexible, highly integrated RFID module.

Carbon Nanotubes (CNT) composites have been found to have electrical conductance highly sensitive to extremely small quantities of gases, such as ammonia (NH_3) and nitrogen oxide (NOx) [21], and be compatible with inkjet-printing [22]. However, due to the insufficient molecular network formation among the inkjet-printed CNT particles at nano-scale, instabilities were observed in both the resistance and, especially, the reactance dependence on frequency above several MHz, which limits the CNT application in only DC or LF band [23]. To enable the CNT-enabled sensor to be integrated with RFID antenna at UHF band, a special recipe needs to be developed.

Two types of SWCNT, namely, P2-SWNT and P3-SWNT were tested. P2-SWNT is developed from purified AP-SWNT by air oxidation and catalyst removing. P3-SWNT is developed from AP-SWNT purified with nitric acid. Compared with P2-SWNT, P3-SWNT has much higher functionality and is easier to disperse in the solvent. In experiments, P2-SWNT started to aggregate at the concentration lower than 0.1 mg/ml, while P3-SWNT can go up to 0.4 mg/ml and still show good dispersion. Therefore, P3-SWNT was selected for latter steps.

The sample SWCNT powder was dispersed in DMF, a polar aprotic solvent. The concentration of the ink was 0.4 mg/ml. This high concentration helped the nano particle network formation after printing; otherwise there would be instability in the impedance response versus frequency of the SWCNT film due to insufficient network formation, such as a sharp dropping of resistance value after 10 MHz [20]. The diluted solution was purified by sonication for 12 hours to prevent aggregations of large particle residues. This is important to avoid the nozzle clogging by SWCNT flocculation during the printing process. Dimatix DMCLCP-11610 printer head was used to eject the SWCNT ink droplet.

Silver electrodes were patterned with the nano-practical ink from Cabot before depositing the SWCNT film, followed by a 140°C sintering. The electrode finger is 2 mm by 10 mm with a gap of 0.8 mm. Then, the 3 mm by 2 mm SWCNT film was deposited. The 0.6 mm overlapping zone is to ensure the good contact between the SWCNT film and the electrodes. Four devices with 10, 15, 20, and 25 SWCNT were fabricated to investigate the electrical properties. Fig. 5.15 shows the fabricated samples.

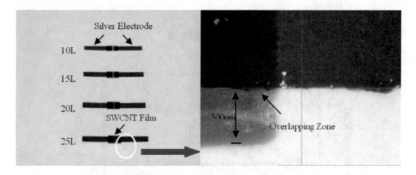

Figure 5.15: Photograph of the inkjet-printed SWCNT films with silver electrodes. The SWCNT layers of the samples from up to down are 10, 15, 20, and 25, respectively.

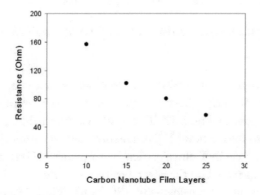

Figure 5.16: Measured DC electrical resistance of SWCNT films.

CNT composites have an affinity for gas molecules. The absorption of gas molecules in the CNT tubes changes the conductivity of the material, which can be explained by the charge transfer of reactive gas molecules with semiconducting CNT. The electrical resistance of the fabricated device was measured by probing the end tips of the two electrodes. The DC results are shown in Fig. 5.16. The resistance goes down when the number of SWCNT layers increases. Since a high number of SWCNT overwritten layers will also help the nano particle network formation, 25-layer film is expected to have the most stable impedance-frequency response and selected for the gas

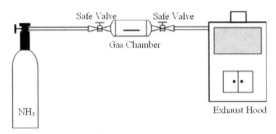

Figure 5.17: Schematic of NH3 gas detection measurement.

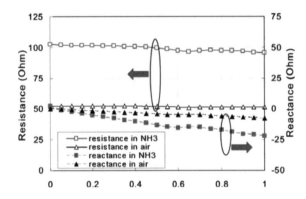

Figure 5.18: Measured impedance characteristics of SWCNT film with 25 layers.

measurement. In the experiment, 4% consistency ammonia, which was widely used in chemistry plants, was guided into an 18-inch tube-shape gas flowing chamber connected with an exhaust hood. The test setup is shown in Fig. 5.17. The SWCNT film exhibits a fast while monotonic impedance response curve to the gas flow [24]. A network vector analyzer (Rohde&Schwarz ZVA8) was used to characterize the SWCNT film electrical performance at UHF band before and after the gas flowing. A GS probe was placed on the silver electrodes for the impedance measurements. The calibration method used was short-open-load-thru (SOLT). In Fig. 5.18, the gas sensor of SWCNT composite shows a very stable impedance response up to 1 GHz, which verifies the effectiveness of the developed SWCNT solvent recipe. At 868 MHz, the sensor exhibits a resistance of 51.6Ω and a reactance of $-j6.1\Omega$ in air. After meeting ammonia, the resistance was increased to 97.1Ω and the reactance was shifted to $-j18.8\Omega$.

A passive RFID system operates in the following way: the RFID reader sends an interrogating RF signal to the RFID tag consisting of an antenna and an IC chip as a load. The IC responds to the reader by varying its input impedance, thus modulating the backscattered signal. The modulation scheme often used in RFID applications is amplitude shift keying (ASK) in which the IC impedance switches between the matched state and the mismatched state [25]. The power reflection coefficient of the RFID antenna can be calculated as a measure to evaluate the reflected wave strength.

$$\eta = \left| \frac{Z_{Load} - Z_{ANT}*}{Z_{Load} + Z_{ANT}} \right|^2 \tag{5.1}$$

where Z_{Load} represents the impedance of the load and Z_{ANT} represents the impedance of the antenna terminals with $Z_{ANT}*$ being its complex conjugate. The same mechanism can be used to realize RFID-enabled sensor modules. The SWCNT film functions as a tunable resistor Z_{Load} with a value determined by the existence of the target gas. The RFID reader monitors the backscattered power level. When the power level changes, it means that there is variation in the load impedance, therefore the sensor detects the existence of the gas, as illustrated in Fig. 5.19.

The expected power levels of the received signal at the load of the RFID antenna can be calculated using Friis free-space formula, as

$$P_{tag} = P_t G_t G_r \left(\frac{\lambda}{4\pi d} \right)^2 \tag{5.2}$$

where P_t is the power fed into the reader antenna, G_t and G_r is the gain of the reader antenna and tag antenna, respectively, and d is the distance between the reader and the tag.

Due to the mismatch between the SWCNT sensor and tag antenna, a portion of the received power would be reflected back, as

$$P_{ref} = P_{tag}\eta \tag{5.3}$$

where η is the power reflection coefficient in (5.1). Hence, the backscattered power received by the RFID reader is defined as

$$P_r = P_{ref} G_t G_r \eta \left(\frac{\lambda}{4\pi d} \right)^2 = P_t G_t^2 G_r^2 \eta \left(\frac{\lambda}{4\pi d} \right)^4 \tag{5.4}$$

or written in a decibel form, as

$$P_r = P_t + 2G_t + 2G_r - 40 \log_{10} \left(\frac{4\pi}{\lambda} \right) - 40 \log_{10} (d) + \eta \tag{5.5}$$

where except the term of η, all the other values remain constant before and after the RFID tag meets gas. Therefore, the variation of the backscattered power level solely depends on η, which is determined by the impedance of the SWCNT film.

A bow-tie meander line dipole antenna was designed and fabricated on a 100μm thickness flexible paper substrate with dielectric constant 3.2. The RFID prototype structure is shown in Fig. 5.20 along with dimensions, with the SWCNT film inkjet printed in the center. The nature of the bow-tie shape offers a more broadband operation for the dipole antenna.

A dielectric probe station was used for the impedance measurements. The measured Z_{ANT} at 868 MHz is 42.6+j11.4Ω. The simulation and measurement results of the return loss of the

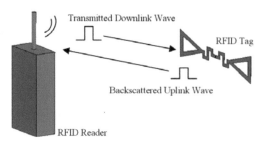

Figure 5.19: Conceptual diagram of the proposed RFID-enabled sensor module.

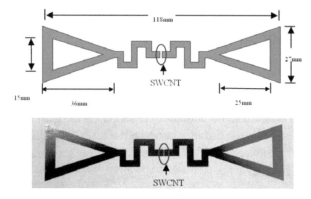

Figure 5.20: The RFID tag module design on flexible substrate: (a) configuration (b) photograph of the tag with inkjet-printed SWCNT film as a load.

proposed antenna are shown in Fig. 5.21, showing a good agreement. The tag bandwidth extends from 810 MHz to 890 MHz, covering the whole European RFID band. The radiation pattern is plotted in Fig. 5.22, which is almost omnidirectional at 868 MHz with directivity around 2.01 dBi and 94.2% radiation efficiency. In order to verify the performance of the conformal antenna, measurements were performed as well by sticking the same tag on a 75 mm radius foam cylinder. As shown in Fig. 5.21, there is almost no frequency shifting observed, with a bandwidth extending from 814 MHz to 891 MHz. The directivity is slightly decreased to 1.99 dBi with 90.3% radiation efficiency. Overall a good performance is still remained with the interested band covered. Fig. 5.23 shows the photograph of the designed conformal tag.

In air, the SWCNT film exhibited an impedance of $51.6-j6.1\Omega$, which results in a power reflection at -18.4 dB. When NH3 is present, SWCNT film's impedance was shifted to $97.1-j18.8\Omega$. The mismatch at the antenna port increased the power reflection to -7.6 dB. From Equation (5.5), there would be 10.8 dBi increase at the received backscattered power level, as shown in Fig. 5.24. By detecting this backscattered power difference on the reader's side, the sensing function can be fulfilled.

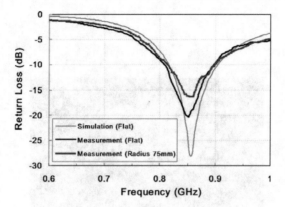

Figure 5.21: Simulated and measured return loss of the RFID tag antenna.

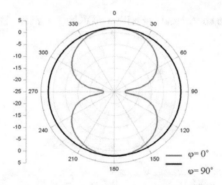

Figure 5.22: Far-field radiation pattern plots.

In this work, the inkjet printing method has been utilized for the first time to deposit SWCNT film on a fully-printed UHF RFID module on paper to form a wireless gas sensor node. To ensure reliable inkjet printing, a SWCNT ink solution has been developed. The impedance performance of the SWCNT film was also characterized up to 1 GHz for the first time. The design demonstrates the great applicability of inkjet-printed CNT for the realization of fully-integrated "green" wireless RFID-enabled flexible sensor nodes based on the ultrasensitive variability of the resistive properties of the CNT materials, and also illustrates the design of novel ultrasensitive passive RFID-enabled sensors.

Figure 5.23: Photograph of the conformal tag with a SWCNT film in the center.

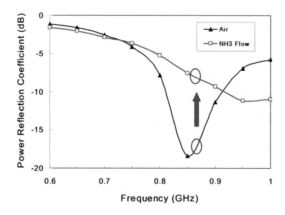

Figure 5.24: The power reflection coefficient of the RFID tag antenna with a SWCNT film before and after the gas flow.

Bibliography

[1] L. Yang, A. Rida, R. Vyas, and M. M. Tentzeris, "RFID tag and RF structures on a paper substrate using inkjet-printing technology," IEEE Transaction on Microwave Theory and Techniques, vol. 55, pp. 2894–2901, Dec. 2007. DOI: 10.1109/TMTT.2007.909886

[2] L. Yang, A. Rida, R. Vyas, and M. M. Tentzeris, "RFID Tag and RF Structures on Paper Substrates using Inkjet-Printing Technology," *IEEE Transactions on Microwave Theory and Techniques*, vol. 55, No.12, Part 2, pp. 2894–2901, December 2007. DOI: 10.1109/TMTT.2007.909886

[3] A. Rida, L. Yang, R. Vyas, S. Bhattacharya, and M. Tentzeris, "Design and integration of inkjet-printed paper-based UHF components for RFID and ubiquitous sensing applications," IEEE European Microwave Conference, pp. 724–727, Oct. 2007. DOI: 10.1109/EUMC.2007.4405294

[4] EPCglobal, "EPC radio-frequency identity protocols Class-1 Generation-2 UHF RFID air interface version 1.0.9," 2005.

[5] R. Vyas, A. Rida, L. Yang, and M. Tentzeris, "Design and Development of a Novel Paper-based Inkjet-Printed RFID-Enabled UHF (433.9 MHz) Sensor Node," IEEE Asia Pacific Microwave Conference, pp. 1–4, Dec. 2007. DOI: 10.1109/APMC.2007.4554641

[6] J. Peatman, Embedded Design with the PIC18F452 Microcontroller. Upper Saddle River, NJ: Pearson, pp. 51–68 & 116-131, 2003.

[7] Panasonic Magnesium Lithium Coin Batteries Specifications. Panasonic, 2005.

[8] "UHF Gen-2 System Overview." Texas Instruments, Sept. 2005.

[9] U. Rohde, Microwave and Wireless Synthesizers: Theory and Design, Paterson, NJ: John Wiley & Sons, pp. 1–5, 1997.

[10] L. Yang, S. Basat, and A. Rida, "Design and development of novel miniaturized UHF RFID tags on ultra-low-cost paper-based substrates", IEEE Asia Pacific Microwave Conference, pp. 1493–1496, Dec. 2006. DOI: 10.1109/APMC.2006.4429689

[11] L. Yang, A. Rida, T. Wu, S. Basat, and M. Tentzeris, "Integration of sensors and inkjet-printed RFID tags on paper-based substrates for UHF "Cognitive Intelligence Applications," IEEE Antennas and Propagation International Symposium, pp. 1193–1196, June 2007. DOI: 10.1109/PIMRC.2007.4394346

[12] C. Balanis, Antenna Theory. New York: Wiley, pp. 162, 133–143, 412–414, 32–34, 1997.

[13] A. Rida, L. Yang, and M. Tentzeris, "Design and characterization of novel paper-based inkjet-printed UHF antennas for RFID and sensing applications," IEEE Antennas and Propagation International Symposium, pp. 2749–2752, June 2007. DOI: 10.1109/APS.2007.4396104

[14] S. Cripps, RF Power Amplifiers for Wireless Communication, Norwood, MA: Artech House, pp. 1–32, 1999.

[15] Radioshack Digital Infrared Thermometer Owner's Manual. Radioshack Corporation, Fort Worth, TX, 2001.

[16] B. Kim, S. Nikolaou, G. E. Ponchak, Y.-S. Kim, J. Papapolymerou, and M. M. Tentzeris, "A Curvature CPW-fed Ultra-wideband Monopole Antenna on Liquid Crystal Polymer Substrate Using Flexible Characteristics." Antenna and Propagation Society International Symposium, pp. 1667–1670, Albuquerque, NM, July 2006. DOI: 10.1109/APS.2006.1710881

[17] Seong H. Lee, Jong K. Park, and Jung N. Lee, "BA novel CPW-fed ultra-wideband antenna design." Microwave Opt. Technol. Lett., vol. 44, no. 5, pp. 393–396, Mar. 2005. DOI: 10.1002/mop.20646

[18] H. W. Ott, Noise Reduction Techniques in Electronic Systems, 2nd ed., Wiley, pp. 73–115, 1988.

[19] M. Philipose, J. Smith, B. Jiang, A. Mamishev, and K. Sundara-Rajan, "Battery-free wireless identification and sensing," IEEE Pervasive Computing, vol. 4, Issue 1, pp. 37–45, 2005. DOI: 10.1109/MPRV.2005.7

[20] S. Johan, Z. Xuezhi, T. Unander, A. Koptyug, and H. Nilsson, "Remote Moisture Sensing utilizing Ordinary RFID Tags," IEEE Sensors 2007, pp. 308–311, 2007. DOI: 10.1109/ICSENS.2007.4388398

[21] K. G. Ong and K. Zeng, "A wireless, passive carbon nanotube-based gas sensor," IEEE Sens. Journal, vol. 2, pp. 82–88, 2002. DOI: 10.1109/JSEN.2002.1000247

[22] J. Song, J. Kim, Y. Yoon, B. Choi, and C. Han, "Inkjet printing of singe-walled carbon nanotubes and electrical characterization of the line pattern," Nanotechnology, vol. 19, 2008. DOI: 10.1088/0957-4484/19/9/095702

[23] M. Dragoman, E. Flahaut, D. Dragoman, M. Ahmad, and R. Plana, "Writing electronic devices on paper with carbon nanotube ink," ArXiv-0901.0362, Jan. 2009.

[24] J. Yun, H. Chang-Soo, J. Kim, J. Song, and Y. Park, "Fabrication of carbon nanotube sensor device by inkjet printing," IEEE Nano Engineered and Molecular Systems Conf., pp. 506–509, 2008. DOI: 10.1109/NEMS.2008.4484382

[25] P. V. Nikitin and K. V. S. Rao, "Performance limitations of passive UHF RFID systems," IEEE Symposium on Antennas and Propagation 2006, pp. 1011–1014, July 2006. DOI: 10.1109/APS.2006.1710704

Printed in the United States
by Baker & Taylor Publisher Services